Lóegaire Humphrey (Ed.)

Draconyx

Lóegaire Humphrey (Ed.)

Draconyx

Genus, Dinosaur, Jurassic, Ornithopod

Claud Press

Imprint

Publisher:
Claud Press is a trademark of
International Book Market Service Ltd., 17 Rue Meldrum, Beau Bassin, 1713-01 Mauritius
Email: info@bookmarketservice.com
Website: www.bookmarketservice.com

Published in 2011

Printed in: U.S.A., U.K., Germany. This book was not produced in Mauritius.

ISBN: 978-613-7-49110-2

Contents

Articles

References

Draconyx

Draconyx	
Scientific classification	
Kingdom:	Animalia
Phylum:	Chordata
Class:	Reptilia
Superorder:	Dinosauria
Order:	†Ornithischia
Suborder:	†Ornithopoda
Family:	†Camptosauridae
Genus:	†*Draconyx* Mateus & Antunes, 2001
Species:	†. ***loureiroi***
Binomial name	
Draconyx loureiroi Mateus & Antunes, 2001	

Draconyx (meaning "dragon claw") is a genus of dinosaur from the Late Jurassic. It was an ornithopod which lived in what is now Portugal. It was found in Lourinhã, in 1991, and described by Octávio Mateus and Miguel Telles Antunes in 2001.

Etymology

The type species, known from only partial remains, is *Draconyx loureiroi.* The generic name is derived from Latin *draco*, "dragon", and Ancient Greek ὄνυξ, *onyx*, "claw". The specific name is in honour of João de Loureiro, a Portuguese jesuit priest, a pioneer in Portuguese palaeontology.

Material

The holotype, **ML 357**, a partial skeleton lacking the skull, consists of two maxillary teeth, three caudal centra, one chevron, a distal epiphysis of right humerus, one manual phalanx, three manual unguals, a distal epiphysis of the right femur, the proximal and distal epiphyses of the tibia and fibula, an astragalus, a calcanaeum, three tarsals, four metatarsals and pedal phalanges. It was in 1991 found at Vale de Frades by Carlos Anunciação of the Museu da Lourinhã, in layers of the Bombarral Unit dating to the Tithonian.

A left femur (ML 434), found near Praia do Caniçal, has been referred to this taxon.

Description

Draconyx is a small bipedal herbivore. Gregory S. Paul in 2010 estimated the length at 3.5 metres, the weight at 150 kilograms.[1]

Systematics

According to Mateus and Antunes (2001), *Draconyx loureiroi* is a member of the Iguanodontia, more specifically the Camptosauridae, based on the maxillary teeth, which have a strong vertical primary ridge on the distal side of the labial crown, and the femur, which is curved and has a prominent lesser trochanter.

References

- Mateus, O. & Antunes, M.T. (2001). "*Draconyx loureiroi*, a new Camptosauridae (Dinosauria: Ornithopoda) from the Late Jurassic of Lourinhã, Portugal". *Annales de Paleontologie*, 87(1): 61-73

Notes

[1] Paul, G.S., 2010, *The Princeton Field Guide to Dinosaurs*, Princeton University Press p. 285

External links

- Mateus, O. & Antunes, M.T. (2001). "*Draconyx loureiroi*, a new Camptosauridae (Dinosauria: Ornithopoda) from the Late Jurassic of Lourinhã, Portugal". *Annales de Paleontologie*, 87(1): 61-73 (http://www.museulourinha.org/pt/Omateus/Papers/MateusAntunes2001Draconyx.pdf.pdf)
- Iguanodontia at Thescelosaurus! (http://www.thescelosaurus.com/iguanodontia.htm)

Genus

In biology, a **genus** (plural: **genera**) is a low-level taxonomic rank used in the biological classification of living and fossil organisms, which is an example of definition by genus and differentia. Genera and higher taxonomic levels such as families are used in biodiversity studies, particularly in fossil studies since species cannot always be confidently identified and genera and families typiclaly have longer stratigraphic ranges than species.[1]

The term comes from Latin genus "descent, family, type, gender",[2] cognate with Greek: *γένος* – *genos*, "race, stock, kin".[3]

The composition of a genus is determined by a taxonomist. The standards for genus classification are not strictly codified, and hence different authorities often produce different classifications for genera. In the hierarchy of the binomial classification system, genus comes above species and below family.

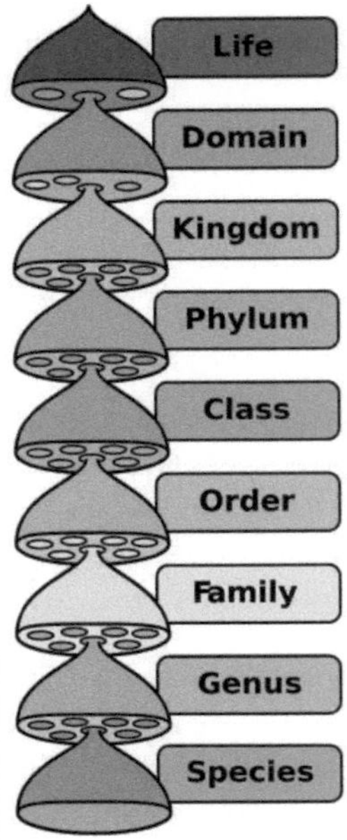

The hierarchy of biological classification's eight major taxonomic ranks, which is an example of definition by genus and differentia. A family contains one or more genera. Intermediate minor rankings are not shown.

Generic name

The scientific name of a genus may be called the **generic name** or **generic epithet**: it is always capitalized. It plays a pivotal role in binomial nomenclature, the system of biological nomenclature.

Binomial nomenclature

The rules for scientific names are laid down in the Nomenclature Codes; depending on the kind of organism and the Kingdom it belongs to, a different Code may apply, with different rules, laid down in a different terminology. The advantages of scientific over common names are that they are accepted by speakers of all languages, and that each species has only one name. This reduces the confusion that may arise from the use of a common name to designate different things in different places (example elk), or from the existence of several common names for a single species.

It is possible for a genus to be assigned to a kingdom governed by one particular Nomenclature Code by one taxonomist, while other taxonomists assign it to a kingdom governed by a different Code, but this is the exception, not the rule.

Pivotal in binomial nomenclature

The generic name often is a component of the names of taxa of lower rank. For example, *Canis lupus* is the scientific name of the Gray wolf, a species, with *Canis* the generic name for the dog and its close relatives, and with *lupus* particular (specific) for the wolf (*lupus* is written in lower case). Similarly, *Canis lupus familiaris* is the scientific name for the domestic dog.

Taxonomic units in higher ranks often have a name that is based on a generic name, such as the family name Canidae, which is based on *Canis*. However, not all names in higher ranks are necessarily based on the name of a genus: for example, Carnivora is the name for the order to which the dog belongs.

The problem of identical names used for different genera

A genus in one kingdom is allowed to bear a scientific name that is in use as a generic name (or the name of a taxon in another rank) in a kingdom that is governed by a different nomenclature code. Although this is discouraged by both the International Code of Zoological Nomenclature and the International Code of Botanical Nomenclature, there are some five thousand such names that are in use in more than one kingdom. For instance, *Anura* is the name of the order of frogs but also is the name of a genus of plants (although not current: it is a synonym); *Aotus* is the genus of golden peas and night monkeys; *Oenanthe* is the genus of wheatears and water dropworts, *Prunella* is the genus of accentors and self-heal, and *Proboscidea* is the order of elephants and the genus of devil's claws.

Within the same kingdom one generic name can apply to only one genus. This explains why the platypus genus is named *Ornithorhynchus*—George Shaw named it *Platypus* in 1799, but the name *Platypus* had already been given to a group of ambrosia beetles by Johann Friedrich Wilhelm Herbst in 1793. Names with the same form but applying to different taxa are called homonyms. Since beetles and platypuses are both members of the kingdom Animalia, the name *Platypus* could not be used for both. Johann Friedrich Blumenbach published the replacement name *Ornithorhynchus* in 1800.

Types and genera

Because of the rules of scientific naming, or "binomial nomenclature", each genus should have a designated type, although in practice there is a backlog of older names that may not yet have a type. In zoology this is the type species; the generic name is permanently associated with the type specimen of its type species. Should this specimen turn out to be assignable to another genus, the generic name linked to it becomes a junior synonym, and the remaining taxa in the former genus need to be reassessed.

See scientific classification and nomenclature codes for more details of this system. Also see type genus.

Guidelines

There are no hard and fast rules that a taxonomist has to follow in deciding what does and what does not belong in a particular genus. This does not mean that there is no common ground among taxonomists in what constitutes a "good" genus. For instance, some rules-of-thumb for delimiting a genus are outlined in Gill.[4] According to these, a genus should fulfill three criteria to be descriptively useful:

1. monophyly – all descendants of an ancestral taxon are grouped together;
2. reasonable compactness – a genus should not be expanded needlessly; and
3. distinctness – in regards of evolutionarily relevant criteria, i.e. ecology, morphology, or biogeography; note that DNA sequences are a *consequence* rather than a *condition* of diverging evolutionary lineages except in cases where they directly inhibit gene flow (e.g. postzygotic barriers).

Nomenclature

...difficulties occurring in generic nomenclature: similar cases abound, and become complicated by the different views taken of the matter by the various taxonomists.

Prof. C. S. Rafinesque. 1836[5]

None of the nomenclature codes require such criteria for defining a genus, because these are concerned with the nomenclature rules, not with taxonomy. These regulate formal nomenclature, aiming for universal and stable scientific names.

See also

- Andrew Delmar Hopkins
- List of the largest genera of flowering plants

References

[1] Sahney, S., Benton, M.J. and Ferry, P.A. (2010). "Links between global taxonomic diversity, ecological diversity and the expansion of vertebrates on land" (http://rsbl.royalsocietypublishing.org/content/6/4/544.full.pdf+html) (PDF). *Biology Letters* **6** (4): 544–547. doi:10.1098/rsbl.2009.1024. PMC 2936204. PMID 20106856. .

[2] Merriam Webster Dictionary (http://www.merriam-webster.com/dictionary/genus)

[3] Genos (http://www.perseus.tufts.edu/cgi-bin/ptext?doc=Perseus:text:1999.04.0057:entry=#21921), Henry George Liddell, Robert Scott, 'A Greek-English Lexicon, *at Perseus*

[4] Gill, F. B., B. Slikas, and F. H. Sheldon. "Phylogeny of titmice (Paridae): II. Species relationships based on sequences of the mitochondrial cytochrome-b gene." Auk 122(1): 121-143, 2005. (Google Scholar) (http://scholar.google.com/scholar?cluster=16219444564703958615&hl=en)

[5] Rafinesque, Prof. C. S. (1836). "Generic Rules" (http://www.us.archive.org/GnuBook/?id=floratellurianaOOrafi#99). *Flora telluriana Pars Prima First Part of the Synoptical Flora Telluriana, Centuries I, II, III, IV. With new Natural Classes, Orders and families: containing*

the 2000 New or revised Genera and Species of Trees, Palms, Shrubs, Vines, Plants, Lilies, Grasses, Ferns, Algas, Fungi, & c. from North and South America, Polynesia, Australia, Asia Europe and Africa, omitted or mistaken by the authors, that were observed or ascertained, described or revised, collected or figured, between 1796 and 1836. (http://www.us.archive.org/GnuBook/?id=floratelluriana00rafi#13). **1**. Philadelphia: H. Probasco. . Retrieved 2009-04-02. "...difficulties occurring in generic nomenclature: similar cases abound, and become complicated by the different views taken of the matter by the various botanists."

External links

- Nomenclator Zoologicus (http://uio.mbl.edu/NomenclatorZoologicus/): Index of all genus and subgenus names in zoological nomenclature from 1758 to 2004.
- Fauna Europaea Database for Taxonomy (http://www.faunaeur.org/full_results.php?id=193482)

rue:Рід (біологія)

Dinosaur

Dinosaurs (from Greek: *δεινός* terrible or potent, and *σαύρα* lizard) are a diverse group of animals that were the dominant terrestrial vertebrates for over 160 million years, from the late Triassic period (about 230 million years ago) until the end of the Cretaceous (about 65 million years ago), when the Cretaceous–Paleogene extinction event led to the extinction of most dinosaur species at the close of the Mesozoic era. The fossil record indicates that birds evolved within theropod dinosaurs during the Jurassic period. Some of them survived the Cretaceous–Paleogene extinction event, including the ancestors of all modern birds. Consequently, in modern classification systems, birds are considered a type of dinosaur—the only group which survived to the present day.[1] [2]

Dinosaurs are a diverse and varied group of animals; birds, at over 9,000 species, are the most diverse group of vertebrate besides perciform fish.[3] Paleontologists have identified over 500 distinct genera[4] and more than 1,000 different species of non-avian dinosaurs.[5] Dinosaurs are represented on every continent by both extant species and fossil remains.[6] Some dinosaurs are herbivorous, others carnivorous. Many dinosaurs, including birds, have been bipedal, though many extinct groups were quadrupedal, and some were able to shift between these body postures. Many species possess elaborate display structures such as horns or crests, and some prehistoric groups developed even more elaborate skeletal modifications such as bony armor. Avian dinosaurs have been the planet's dominant flying vertebrate since the extinction of the pterosaurs, and evidence suggests that all ancient dinosaurs built nests and laid eggs much as avian species do today. Although generally known for the large size of some species, most Mesozoic dinosaurs were human-sized or even smaller.

The term "dinosaur" was coined in 1842 by the English paleontologist Richard Owen, and derives from Greek *δεινός* (*deinos*) "terrible, powerful, wondrous, potent" + *σαῦρος* (*sauros*) "lizard". Through the first half of the 20th century, most of the scientific community believed dinosaurs to have been sluggish, unintelligent cold-blooded animals. Most research conducted since the 1970s, however, has indicated that dinosaurs were active animals with elevated metabolisms and numerous adaptations for social interaction.

Since the first dinosaur fossils were recognized in the early 19th century, mounted fossil dinosaur skeletons or replicas have been major attractions at museums around the world, and dinosaurs have become a part of world culture. They have been featured in best-selling books and films such as *Jurassic Park*, and new discoveries are regularly covered by the media. In informal speech, the word "dinosaur" is used to describe things that are impractically large, obsolete, or bound for extinction,[7] reflecting the outdated view that dinosaurs were maladapted monsters of the ancient world.

Name

The taxon **Dinosauria** was formally named in 1842 by Sir Richard Owen, who used it to refer to the "distinct tribe or sub-order of Saurian Reptiles" that were then being recognized in England and around the world.[8] :103 The term is derived from the Greek words δεινός (*deinos* meaning "terrible", "powerful", or "wondrous") and σαῦρος (*sauros* meaning "lizard" or "reptile").[8] :103[9] Though the taxonomic name has often been interpreted as a reference to dinosaurs' teeth, claws, and other fearsome characteristics, Owen intended it merely to evoke their size and majesty.[10]

Modern definition

Triceratops skeleton at the American Museum of Natural History in New York City

Formal definitions are written to correspond with scientific conceptions of dinosaurs that predate the modern use of phylogenetics. The continuity of meaning is intended to prevent confusion about what the term "dinosaur" means.

Under phylogenetic taxonomy, dinosaurs are usually defined as the group consisting of "*Triceratops*, Neornithes [modern birds], their most recent common ancestor, and all descendants".[11] It has also been suggested that Dinosauria be defined with respect to the MRCA of *Megalosaurus* and *Iguanodon*, because these were two of the three genera cited by Richard Owen when he recognized the Dinosauria.[12] Both definitions result in the same set of animals being defined as dinosaurs, that is "Dinosauria = Ornithischia + Saurischia", which encompasses theropods (mostly bipedal carnivores and birds), ankylosaurians (armored herbivorous quadrupeds), stegosaurians (plated herbivorous quadrupeds), ceratopsians (herbivorous quadrupeds with horns and frills), ornithopods (bipedal or quadrupedal herbivores including "duck-bills"), and presumptuously, sauropodomorphs (mostly large herbivorous quadrupeds with long necks and tails).

Many paleontologists note that the order in which sauropodomorphs and theropods diverged may omit sauropodomorphs from the definition for both saurischians and dinosaurs. To avoid the instability of Dinosauria, a more conservative definition of Dinosauria is defined with respect to four anchoring nodes: *Triceratops horridus*, *Saltasaurus loricatus*, and *Passer domesticus*, their most recent common ancestor, and all descendants. This "safer" definition can be expressed as "Dinosauria = Ornithischia + Sauropodomorpha + Theropoda".[13]

There is a wide consensus among paleontologists that birds are the descendants of theropod dinosaurs. Using the strict phylogenetic nomenclatural definition that all descendants of a single common ancestor must be included in a group for that group to be natural, birds would thus *be* dinosaurs and dinosaurs are, therefore, not extinct. Birds are classified by most paleontologists as belonging to the subgroup Maniraptora, which are coelurosaurs, which are theropods, which are saurischians, which are dinosaurs.[14]

From the point of view of cladistics, birds are dinosaurs, but in ordinary speech the word "dinosaur" does not include birds. Additionally, referring to dinosaurs that are not birds as "non-avian dinosaurs" is cumbersome. For clarity, this article will use "dinosaur" as a synonym for "non-avian dinosaur". The term "non-avian dinosaur" will be used for emphasis as needed.

General description

Stegosaurus skeleton, Field Museum, Chicago

Using one of the above definitions, dinosaurs (aside from birds) can be generally described as terrestrial archosaurian reptiles with limbs held erect beneath the body, that existed from the Late Triassic (first appearing in the Carnian faunal stage) to the Late Cretaceous (going extinct at the end of the Maastrichtian).[15] Many prehistoric animals are popularly conceived of as dinosaurs, such as ichthyosaurs, mosasaurs, plesiosaurs, pterosaurs, and *Dimetrodon*, but are not classified scientifically as dinosaurs. Marine reptiles like ichthyosaurs, mosasaurs, and plesiosaurs were neither terrestrial nor archosaurs; pterosaurs were archosaurs but not terrestrial; and *Dimetrodon* was a Permian animal more closely related to mammals.[16] Dinosaurs were the dominant terrestrial vertebrates of the Mesozoic, especially the Jurassic and Cretaceous. Other groups of animals were restricted in size and niches; mammals, for example, rarely exceeded the size of a cat, and were generally rodent-sized carnivores of small prey.[17] One notable exception is *Repenomamus giganticus*, a triconodont weighing between 12 kilograms (26 lb) and 14 kilograms (31 lb) that is known to have eaten small dinosaurs like young *Psittacosaurus*.[18]

Dinosaurs were an extremely varied group of animals; according to a 2006 study, over 500 dinosaur genera have been identified with certainty so far, and the total number of genera preserved in the fossil record has been estimated at around 1850, nearly 75% of which remain to be discovered.[4] An earlier study predicted that about 3400 dinosaur genera existed, including many which would not have been preserved in the fossil record.[19] As of September 17, 2008, 1047 different species of dinosaurs have been named.[5] Some were herbivorous, others carnivorous. Some dinosaurs were bipeds, some were quadrupeds, and others, such as *Ammosaurus* and *Iguanodon*, could walk just as easily on two or four legs. Many had bony armor, or cranial modifications like horns and crests. Although known for large size, many dinosaurs were human-sized or smaller. Dinosaur remains have been found on every continent on Earth, including Antarctica.[6] No non-avian dinosaurs are known to have lived in marine habitats or in aerial habitats, although it is possible some feathered non-avian theropods were flyers. There is also evidence that some spinosaurids had semi-aquatic habits.[20]

Distinguishing anatomical features

While recent discoveries have made it more difficult to present a universally agreed-upon list of dinosaurs' distinguishing features, nearly all dinosaurs discovered so far share certain modifications to the ancestral archosaurian skeleton. Although some later groups of dinosaurs featured further modified versions of these traits, they are considered typical across Dinosauria; the earliest dinosaurs had them and passed them on to all their descendants. Such common features across a taxonomic group are called synapomorphies.

A detailed assessment of archosaur interrelations by S. Nesbitt[21] confirmed or found the following 12 unambiguous synapomorphies, some previously known:

- in the skull, a supratemporal fossa (excavation) is present in front of the supratemporal fenestra
- epipophyses present in anterior neck vertebrae (except atlas and axis)
- apex of deltopectoral crest (a projection on which the deltopectoral muscles attach) located at or more than 30% down the length of the humerus (upper arm bone)
- radius shorter than 80% of humerus length
- fourth trochanter (projection where the caudofemoralis muscle attaches) on the femur (thigh bone) is a sharp flange
- fourth trochanter asymmetrical, with distal margin forming a steeper angle to the shaft

- on the astragalus and calcaneum the proximal articular facet for fibula occupies less than 30% of the transverse width of the element
- exocciptials (bones at the back of the skull) do not meet along the midline on the floor of the endocranial cavity
- proximal articular surfaces of the ischium with the ilium and the pubis separated by a large concave surface
- cnemial crest on the tibia (shinbone) arcs anterolaterally
- distinct proximodistally oriented ridge present on the posterior face of the distal end of the tibia

Nesbitt found a number of further potential synapomorphies, and discounted a number of synapomorphies previously suggested. Some of these are also present in silesaurids, which Nesbitt recovered as a sister group to Dinosauria, including a large anterior trochanter, metatarsals II and IV of subequal length, reduced contact between ischium and pubis, the presence of a cenmial crest on the tibia and of an ascending process on the astragalus,[11] and many others.

Edmontonia was an armored dinosaur of the group Ankylosauria.

A variety of other skeletal features were shared by many dinosaurs. However, because they were either common to other groups of archosaurs or were not present in all early dinosaurs, these features are not considered to be synapomorphies. For example, as diapsid reptiles, dinosaurs ancestrally had two pairs of temporal fenestrae (openings in the skull behind the eyes), and as members of the diapsid group Archosauria, had additional openings in the snout and lower jaw.[22] Additionally, several characteristics once thought to be synapomorphies are now known to have appeared before dinosaurs, or were absent in the earliest dinosaurs and independently evolved by different dinosaur groups. These include an elongated scapula, or shoulder blade; a sacrum composed of three or more fused vertebrae (three are found in some other archosaurs, but only two are found in *Herrerasaurus*);[11] and an acetabulum, or hip socket, with a hole at the center of its inside surface (closed in *Saturnalia*, for example).[23] Another difficulty of determining distinctly dinosaurian features is that early dinosaurs and other archosaurs from the Late Triassic are often poorly known and were similar in many ways; these animals have sometimes been misidentified in the literature.[24]

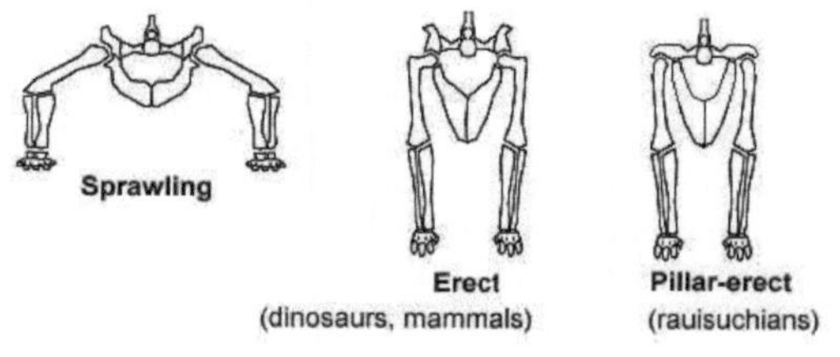

Hip joints and hindlimb postures

Dinosaurs stood erect in a manner similar to most modern mammals, but distinct from most other reptiles, whose limbs sprawl out to either side.[25] Their posture was due to the development of a laterally facing recess in the pelvis (usually an open socket) and a corresponding inwardly facing distinct head on the femur.[26] Their erect posture enabled dinosaurs to breathe easily while moving, which likely permitted stamina and activity levels that surpassed those of "sprawling" reptiles.[27] Erect limbs probably also helped support the evolution of large size by reducing bending stresses on limbs.[28] Some non-dinosaurian archosaurs, including rauisuchians, also had erect limbs but achieved this by a "pillar erect" configuration of the hip joint, where instead of having a projection from the femur insert on a socket on the hip, the upper pelvic bone was rotated to form an overhanging shelf.[28]

Natural history

Origins and early evolution

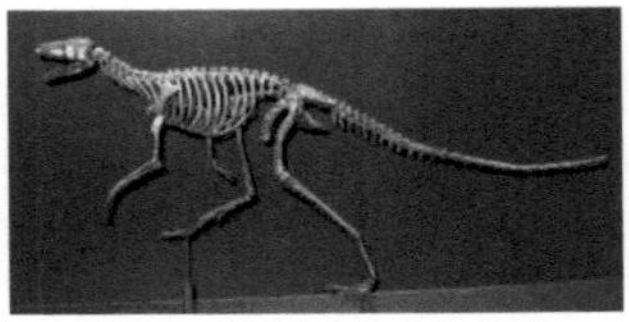
Marasuchus, a dinosaur-like ornithodiran

For a long time many scientists thought dinosaurs were polyphyletic with multiple groups of unrelated "dinosaurs" evolving due to similar pressures,[29] [30] [31] but dinosaurs are now known to have formed a single group.[11] [21] [32] Dinosaurs diverged from their archosaur ancestors approximately 230 million years ago during the Middle to Late Triassic period, roughly 20 million years after the Permian–Triassic extinction event wiped out an estimated 95% of all life on Earth.[33] [34] Radiometric dating of the rock formation that contained fossils from the early dinosaur genus *Eoraptor* establishes its presence in the fossil record at this time. Paleontologists believe *Eoraptor* resembles the common ancestor of all dinosaurs;[35] if this is true, its traits suggest that the first dinosaurs were small, bipedal predators.[36] The discovery of primitive, dinosaur-like ornithodirans such as *Marasuchus* and *Lagerpeton* in Argentinian Middle Triassic strata supports this view; analysis of recovered fossils suggests that these animals were indeed small, bipedal predators.

When dinosaurs appeared, terrestrial habitats were occupied by various types of basal archosaurs and therapsids, such as aetosaurs, cynodonts, dicynodonts, ornithosuchids, rauisuchias, and rhynchosaurs. Most of these other animals became extinct in the Triassic, in one of two events. First, at about the boundary between the Carnian and Norian faunal stages (about 215 million years ago), dicynodonts and a variety of basal archosauromorphs, including the prolacertiforms and rhynchosaurs, became extinct. This was followed by the Triassic–Jurassic extinction event (about 200 million years ago), that saw the end of most of the other groups of early archosaurs, like aetosaurs, ornithosuchids, phytosaurs, and rauisuchians. These losses left behind a land fauna of crocodylomorphs, dinosaurs, mammals, pterosaurians, and turtles.[11]

The early forms *Herrerasaurus* (large), *Eoraptor* (small) and a *Plateosaurus* skull

The first few lines of primitive dinosaurs diversified through the Carnian and Norian stages of the Triassic, most likely by occupying the niches of groups that became extinct. Traditionally, dinosaurs were thought to have replaced the variety of other Triassic land animals by proving superior through a long period of competition. This now appears unlikely, for several reasons. Dinosaurs do not show a pattern of steadily increasing in diversity and numbers, as would be predicted if they were competitively replacing other groups; instead, they were very rare through the Carnian, making up only 1–2% of individuals present in faunas. In the Norian, however, after the extinction of several other groups, they became significant components of faunas, representing 50–90% of individuals. Also, what had been viewed as a key adaptation of dinosaurs, their erect stance, is now known to have been present in several contemporaneous groups that were not as successful (aetosaurs, ornithosuchids, rauisuchians, and some groups of crocodylomorphs). Finally, the Late Triassic itself was a time of great upheaval in life, with shifts in plant life, marine life, and climate.[11] Crurotarsans, today represented only by crocodilians but in the Late Triassic also encompassing such now-extinct groups as aetosaurs, phytosaurs, ornithosuchians, and rauisuchians, were actually more diverse in the Late Triassic than dinosaurs, indicating that the survival of dinosaurs had more to do with luck than superiority.[37]

Classification

Dinosaurs (including birds) are archosaurs, like modern crocodilians. Archosaurs' diapsid skulls have two holes, called temporal fenestrae, located where the jaw muscles attach, and an additional antorbital fenestra in front of the eyes. Most reptiles (including birds) are diapsids; mammals, with only one temporal fenestra, are called synapsids; and turtles, with no temporal fenestra, are anapsids. Anatomically, dinosaurs share many other archosaur characteristics, including teeth that grow from sockets rather than as direct extensions of the jawbones. Within the archosaur group, dinosaurs are differentiated most noticeably by their gait. Dinosaur legs extend directly beneath the body, whereas the legs of lizards and crocodilians sprawl out to either side.

Collectively, dinosaurs are usually regarded as a superorder or an unranked clade. They are divided into two orders, Saurischia and Ornithischia, depending upon pelvic structure. Saurischia includes those taxa sharing a more recent common ancestor with birds than with Ornithischia, while Ornithischia includes all taxa sharing a more recent common ancestor with *Triceratops* than with Saurischia. Saurischians ("lizard-hipped", from the Greek *sauros* (*σαυρος*) meaning "lizard" and *ischion* (*ισχιον*) meaning "hip joint") retained the hip structure of their ancestors, with a pubis bone directed cranially, or forward.[26] This basic form was modified by rotating the pubis backward to varying degrees in several groups (*Herrerasaurus*,[38] therizinosauroids,[39] dromaeosaurids,[40] and birds[14]). Saurischia includes the theropods (bipedal and mostly carnivores, except for birds) and sauropodomorphs (long-necked quadrupedal herbivores).

By contrast, ornithischians ("bird-hipped", from the Greek *ornitheios* (*ορνιθειος*) meaning "of a bird" and *ischion* (*ισχιον*) meaning "hip joint") had a pelvis that superficially resembled a bird's pelvis: the pubis bone was oriented caudally (rear-pointing) Unlike birds, the ornithischian pubis also usually had an additional forward-pointing process. Ornithischia includes a variety of herbivores. (**NB:** the terms "lizard hip" and "bird hip" are misnomers – birds evolved from dinosaurs with "lizard hips".)

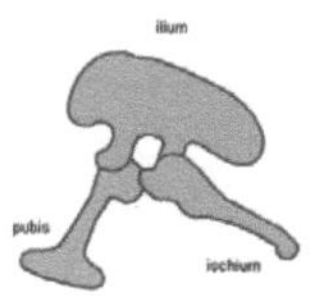

Saurischian pelvis structure (left side)

Tyrannosaurus pelvis (showing saurischian structure – left side)

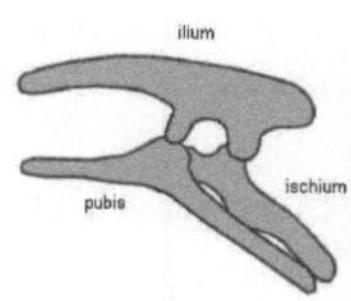

Ornithischian pelvis structure (left side)

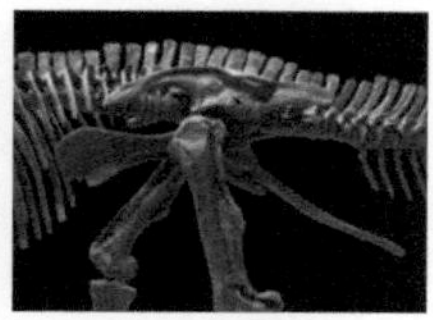

Edmontosaurus pelvis (showing ornithischian structure – left side)

The following is a simplified classification of dinosaur families. A more detailed version can be found at List of dinosaur classifications.

Several macronarian Sauropods: from left to right *Camarasaurus*, *Brachiosaurus*, *Giraffatitan*, and *Euhelopus*

- Dinosauria
  - Saurischia (theropods and sauropods)
    - †Herrerasaurians (early bipedal predators)
    - Theropods (all bipedal; most were carnivores)
      - †Coelophysoids (*Coelophysis* and close relatives)
      - †Ceratosaurians (*Ceratosaurus* and abelisaurids – the latter were important Late Cretaceous predators in southern continents)
      - †Spinosauroids (long bodies; short arms; some with crocodile-like skulls and bony "sails" on their backs)

Various ornithopod dinosaurs and one heterodontosaurid. Far left: *Camptosaurus*, left: *Iguanodon*, center background: *Shantungosaurus*, center foreground: *Dryosaurus*, right: *Corythosaurus*, far right (small): *Heterodontosaurus*, far right (large) *Tenontosaurus*.

- †Carnosaurians (*Allosaurus* and close relatives, like *Carcharodontosaurus*)
- Coelurosaurians (diverse, with a range of body sizes and niches)
  - †Tyrannosauroids (small to gigantic, often with reduced forelimbs)
  - †Ornithomimosaurians ("ostrich-mimics"; mostly toothless; carnivores to possible herbivores)
  - †Therizinosauroids (bipedal herbivores with large hand claws and small heads)
  - †Oviraptorosaurians (mostly toothless; their diet and lifestyle are uncertain)
  - †Dromaeosaurids (popularly known as "raptors"; bird-like carnivores)
  - †Troodontids (similar to dromaeosaurids, but more lightly built)
  - Avialans (flying dinosaurs, including modern birds: the only living dinosaurs)

- †Sauropodomorphs (quadrupedal herbivores with small heads, long necks and tails, and elephant-like bodies)
  - †"Prosauropods" (early relatives of sauropods; small to quite large; some possibly omnivorous; bipeds and quadrupeds)
  - †Sauropods (very large, usually over 15 meters long [49 ft])
    - †Diplodocoids (skulls and tails elongated; teeth typically narrow and pencil-like)
    - †Macronarians (boxy skulls; spoon- or pencil-shaped teeth)
      - †Brachiosaurids (very long necks; forelimbs longer than hindlimbs)
      - †Titanosaurians (diverse; stocky, with wide hips; most common in the Late Cretaceous of southern continents)

- †Ornithischians (diverse bipedal and quadrupedal herbivores)
  - †Heterodontosaurids (meter- or yard-scale herbivores or omnivores with prominent canine teeth)
  - †Thyreophorans (armored dinosaurs; mostly quadrupeds)
    - †Ankylosaurians (scutes as primary armor; some had club-like tails)
    - †Stegosaurians (spikes and plates as primary armor)
  - †Ornithopods (diverse, from meter- or yard-scale bipeds to 12-meter (39 ft) animals that could move as both bipeds and quadrupeds; evolved a method of chewing using skull flexibility and large numbers of teeth)
    - †Hadrosaurids ("duckbilled dinosaurs")
  - †Pachycephalosaurians ("bone-heads"; bipeds with domed or knobby growth on skulls)
  - †Ceratopsians (dinosaurs with horns and frills, although most early forms had only the beginnings of these features)

Evolution and paleobiogeography

Dinosaur evolution after the Triassic follows changes in vegetation and the location of continents. In the Late Triassic and Early Jurassic, the continents were connected as the single landmass Pangaea, and there was a worldwide dinosaur fauna mostly composed of coelophysoid carnivores and prosauropod herbivores.[41] Gymnosperm plants (particularly conifers), a potential food source, radiated in the Late Triassic. Prosauropods did not have sophisticated mechanisms for processing food in the mouth, and so must have employed other means of breaking down food farther along the digestive tract.[42] The general homogeneity of dinosaurian faunas continued into the Middle and Late Jurassic, where most localities had predators consisting of ceratosaurians, spinosauroids, and carnosaurians, and herbivores consisting of stegosaurian ornithischians and large sauropods. Examples of this include the Morrison Formation of North America and Tendaguru Beds of Tanzania. Dinosaurs in China show some differences, with specialized sinraptorid theropods and unusual, long-necked sauropods like *Mamenchisaurus*.[41] Ankylosaurians and ornithopods were also becoming more common, but prosauropods had become extinct. Conifers and pteridophytes were the most common plants. Sauropods, like the earlier prosauropods, were not oral processors, but ornithischians were evolving various means of dealing with food in the mouth, including potential cheek-like organs to keep food in the mouth, and jaw motions to grind food.[42] Another notable evolutionary event of the Jurassic was the appearance of true birds, descended from maniraptoran coelurosaurians.[14]

By the Early Cretaceous and the ongoing breakup of Pangaea, dinosaurs were becoming strongly differentiated by landmass. The earliest part of this time saw the spread of ankylosaurians, iguanodontians, and brachiosaurids through Europe, North America, and northern Africa. These were later supplemented or replaced in Africa by large spinosaurid and carcharodontosaurid theropods, and rebbachisaurid and titanosaurian sauropods, also found in South America. In Asia, maniraptoran coelurosaurians like dromaeosaurids, troodontids, and oviraptorosaurians became the common theropods, and ankylosaurids and early ceratopsians like *Psittacosaurus* became important herbivores. Meanwhile, Australia was home to a fauna of basal ankylosaurians, hypsilophodonts, and iguanodontians.[41] The stegosaurians appear to have gone extinct at some point in the late Early Cretaceous or early Late Cretaceous. A major change in the Early Cretaceous, which would be amplified in the Late Cretaceous, was the evolution of flowering plants. At the same time, several groups of dinosaurian herbivores evolved more sophisticated ways to orally process food. Ceratopsians developed a method of slicing with teeth stacked on each other in batteries, and iguanodontians refined a method of grinding with tooth batteries, taken to its extreme in hadrosaurids.[42] Some sauropods also evolved tooth batteries, best exemplified by the rebbachisaurid *Nigersaurus*.[43]

There were three general dinosaur faunas in the Late Cretaceous. In the northern continents of North America and Asia, the major theropods were tyrannosaurids and various types of smaller maniraptoran theropods, with a predominantly ornithischian herbivore assemblage of hadrosaurids, ceratopsians, ankylosaurids, and pachycephalosaurians. In the southern continents that had made up the now-splitting Gondwana, abelisaurids were the common theropods, and titanosaurian sauropods the common herbivores. Finally, in Europe, dromaeosaurids, rhabdodontid iguanodontians, nodosaurid ankylosaurians, and titanosaurian sauropods were prevalent.[41] Flowering plants were greatly radiating,[42] with the first grasses appearing by the end of the Cretaceous.[44] Grinding hadrosaurids and shearing ceratopsians became extremely diverse across North America and Asia. Theropods were also radiating as herbivores or omnivores, with therizinosaurians and ornithomimosaurians becoming common.[42]

The Cretaceous–Paleogene extinction event, which occurred approximately 65 million years ago at the end of the Cretaceous period, caused the extinction of all dinosaurs except for the birds. Some other diapsid groups, such as crocodilians, lizards, snakes, sphenodontians, and choristoderans, also survived the event.[45]

Paleobiology

Knowledge about dinosaurs is derived from a variety of fossil and non-fossil records, including fossilized bones, feces, trackways, gastroliths, feathers, impressions of skin, internal organs and soft tissues.[46] [47] Many fields of study contribute to our understanding of dinosaurs, including physics (especially biomechanics), chemistry, biology, and the earth sciences (of which paleontology is a sub-discipline). Two topics of particular interest and study have been dinosaur size and behavior.

Size

Scale diagram comparing the largest known dinosaurs in five major clades and a human

The sauropods were the largest dinosaurs. For much of the dinosaur era, the smallest sauropods were larger than anything else in their habitat, and the largest were an order of magnitude more massive than anything else that has since walked the Earth. Giant prehistoric mammals such as the *Paraceratherium* and the Columbian mammoth were dwarfed by the giant sauropods, and only a handful of modern aquatic animals approach or surpass them in size – most notably the blue whale, which reaches up to 173000 kg (381000 lb) and over 30 meters (98 ft) in length. There are several proposed advantages for the large size of sauropods, including protection from predation, reduction of energy use, and longevity, but it may be that the most important advantage was dietary. Large animals are more efficient at digestion than small animals, because food spends more time in their digestive systems. This also permits them to subsist on food with lower nutritive value than smaller animals. Sauropod remains are mostly found in rock formations interpreted as dry or seasonally dry, and the ability to eat large quantities of low-nutrient browse would have been advantageous in such environments.[48]

Most dinosaurs, however, were much smaller than the giant sauropods. Current evidence suggests that dinosaur average size varied through the Triassic, early Jurassic, late Jurassic and Cretaceous periods.[35] Theropod dinosaurs, when sorted by estimated weight into categories based on order of magnitude, most often fall into the 100 to 1000 kilogram (220 to 2200 lb) category, whereas recent predatory carnivorans peak in the 10 to 100 kilogram (22 to 220 lb) category.[49] The mode of dinosaur body masses is between one and ten metric tonnes.[50] This contrasts sharply with the size of Cenozoic mammals, estimated by the National Museum of Natural History as about 2 to 5 kilograms (5 to 10 lb).[51]

Largest and smallest

Only a tiny percentage of animals ever fossilize, and most of these remain buried in the earth. Few of the specimens that are recovered are complete skeletons, and impressions of skin and other soft tissues are rare. Rebuilding a complete skeleton by comparing the size and morphology of bones to those of similar, better-known species is an inexact art, and reconstructing the muscles and other organs of the living animal is, at best, a process of educated guesswork. As a result, scientists will probably never be certain of the largest and smallest dinosaurs.

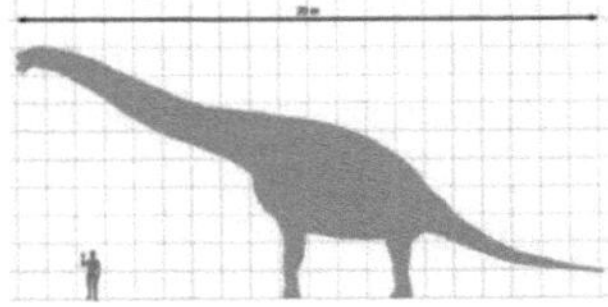

Comparative size of *Giraffatitan*

The tallest and heaviest dinosaur known from good skeletons is *Giraffatitan brancai* (previously classified as a species of *Brachiosaurus*). Its remains were discovered in Tanzania between 1907–12. Bones from several similar-sized individuals were incorporated into the skeleton now mounted and on display at the Museum für Naturkunde Berlin;[52] this mount is 12 meters (39 ft) tall and 22.5 meters (74 ft) long, and would have belonged to an animal that weighed between 30000 and 60000 kilograms (70000 and 130000 lb). The longest complete dinosaur is the 27-meter (89 ft) long *Diplodocus*, which was discovered in Wyoming in the United States and displayed in Pittsburgh's Carnegie Natural History Museum in 1907.

Comparative size of *Eoraptor*

There were larger dinosaurs, but knowledge of them is based entirely on a small number of fragmentary fossils. Most of the largest herbivorous specimens on record were all discovered in the 1970s or later, and include the massive *Argentinosaurus*, which may have weighed 80000 to 100000 kilograms (90 to 110 short tons); some of the longest were the 33.5 meters (110 ft) long *Diplodocus hallorum*[48] (formerly *Seismosaurus*) and the 33 meters (108 ft) long *Supersaurus*;[53] and the tallest, the 18 meters (59 ft) tall *Sauroposeidon*, which could have reached a sixth-floor window. The heaviest and longest of them all may have been *Amphicoelias fragillimus*, known only from a now lost partial vertebral neural arch described in 1878. Extrapolating from the illustration of this bone, the animal may have been 58 meters (190 ft) long and weighed over 120000 kg (260000 lb).[48] The largest known carnivorous dinosaur was *Spinosaurus*, reaching a length of 16 to 18 meters (50 to 60 ft), and weighing in at 8150 kg (18000 lb).[54] Other large meat-eaters included *Giganotosaurus*, *Carcharodontosaurus* and *Tyrannosaurus*.[55]

Not including modern birds, the smallest known dinosaurs known were about the size of a pigeon.[56] The theropods *Anchiornis* and *Epidexipteryx* both had a total skeletal length of under 35 centimeters (1.1 ft).[56] [57] *Anchiornis* is currently the smallest dinosaur described from an adult specimen, with an estimated weight of 110 grams.[57] The smallest herbivorous dinosaurs included *Microceratus* and *Wannanosaurus*, at about 60 cm (2 ft) long each.[58] [59]

Behavior

Interpretations of dinosaur behavior are generally based on the pose of body fossils and their habitat, computer simulations of their biomechanics, and comparisons with modern animals in similar ecological niches. As such, the current understanding of dinosaur behavior relies on speculation, and will likely remain controversial for the foreseeable future. However, there is general agreement that some behaviors which are common in crocodiles and birds, dinosaurs' closest living relatives, were also common among dinosaurs.

A nesting ground of *Maiasaura* was discovered in 1978.

The first potential evidence of herding behavior was the 1878 discovery of 31 *Iguanodon* dinosaurs which were then thought to have perished together in Bernissart, Belgium, after they fell into a deep, flooded sinkhole and drowned.[60] Other mass-death sites have been subsequently discovered. Those, along with multiple trackways, suggest that gregarious behavior was common in many dinosaur species. Trackways of hundreds or even thousands of herbivores indicate that duck-bills (hadrosaurids) may have moved in great herds, like the American Bison or the African Springbok. Sauropod tracks document that these animals traveled in groups composed of several different species, at least in Oxfordshire, England,[61] although there is not evidence for specific herd structures.[62] Dinosaurs may have congregated in herds for defense, for migratory purposes, or to provide protection for their young. There is evidence that many types of dinosaurs, including various theropods, sauropods, ankylosaurians, ornithopods, and ceratopsians, formed aggregations of immature individuals. One example is a site in Inner Mongolia that has yielded the remains of over 20 *Sinornithomimus*, from one to seven years old. This assemblage is interpreted as a social group that was trapped in mud.[63] The interpretation of dinosaurs as gregarious has also extended to depicting carnivorous theropods as pack hunters working together to bring down large prey.[64] [65] However, this lifestyle is uncommon among the modern relatives of dinosaurs (crocodiles and other reptiles, and birds – Harris's Hawk is a well-documented exception), and the taphonomic evidence suggesting pack hunting in such theropods as *Deinonychus* and *Allosaurus* can also be interpreted as the results of fatal disputes between feeding animals, as is seen in many modern diapsid predators.[66]

Fossilized egg of the oviraptorid *Citipati*, American Museum of Natural History

Jack Horner's 1978 discovery of a *Maiasaura* ("good mother dinosaur") nesting ground in Montana demonstrated that parental care continued long after birth among the ornithopods.[67] There is also evidence that other Cretaceous-era dinosaurs, like Patagonian titanosaurian sauropods (1997 discovery), also nested in large groups.[68] The Mongolian oviraptorid *Citipati* was discovered in a chicken-like brooding position in 1993, which may mean it was covered with an insulating layer of feathers that kept the eggs warm.[69] Parental care is also implied by other finds. For example, the fossilized remains of a grouping of *Psittacosaurus* has been found, consisting of one adult and 34 juveniles; in this case, the large number of juveniles may be due to communal nesting.[70] Additionally, a dinosaur embryo (pertaining to the prosauropod *Massospondylus*) was found without teeth, indicating that some parental care was required to feed the young dinosaur.[71] Trackways have also confirmed parental behavior among ornithopods from the Isle of Skye in northwestern Scotland.[72] Nests and eggs have been found for most major groups of dinosaurs, and it appears likely that dinosaurs communicated with their young, in a manner similar to modern birds and crocodiles.

Artist's rendering of two *Centrosaurus*, herbivorous ceratopsid dinosaurs from the late Cretaceous fauna of North America

The crests and frills of some dinosaurs, like the marginocephalians, theropods and lambeosaurines, may have been too fragile to be used for active defense, and so they were likely used for sexual or aggressive displays, though little is known about dinosaur mating and territorialism. Head wounds from bites suggest that theropods, at least, engaged in active aggressive confrontations.[73]

From a behavioral standpoint, one of the most valuable dinosaur fossils was discovered in the Gobi Desert in 1971. It included a *Velociraptor* attacking a *Protoceratops*,[74] providing evidence that dinosaurs did indeed attack each other.[75] Additional evidence for attacking live prey is the partially healed tail of an *Edmontosaurus*, a hadrosaurid dinosaur; the tail is damaged in such a way that shows the animal was bitten by a tyrannosaur but survived.[75] Cannibalism amongst some species of dinosaurs was confirmed by tooth marks found in Madagascar in 2003, involving the theropod *Majungasaurus*.[76]

Comparisons between the scleral rings of dinosaurs and modern birds and reptiles have been used to infer daily activity patterns of dinosaurs. Although it has been suggested that most dinosaurs were active during the day, these comparisons have shown that small predatory dinosaurs such as dromaeosaurids, *Juravenator*, and *Megapnosaurus* were likely nocturnal. Large and medium-sized herbivorous and omnivorous dinosaurs such as ceratopsians, sauropodomorphs, hadrosaurids, ornithomimosaurs may have been cathemeral, active during short intervals throughout the day, although the small ornithischian *Agilisaurus* was inferred to be diurnal.[77]

Based on current fossil evidence from dinosaurs such as *Oryctodromeus*, some herbivorous species seem to have led a partially fossorial (burrowing) lifestyle,[78] and some bird-like species may have been arboreal (tree climbing), most notably primitive dromaeosaurids such as *Microraptor*[79] and the enigmatic scansoriopterygids.[80] However, most dinosaurs seem to have relied on land-based locomotion. A good understanding of how dinosaurs moved on the ground is key to models of dinosaur behavior; the science of biomechanics, in particular, has provided significant insight in this area. For example, studies of the forces exerted by muscles and gravity on dinosaurs' skeletal structure have investigated how fast dinosaurs could run,[81] whether diplodocids could create sonic booms via whip-like tail snapping,[82] and whether sauropods could float.[83]

Communication and vocalization

The nature of dinosaur communication remains enigmatic, and is an active area of research. In 2008, paleontologist Phil Senter examined the evidence for vocalization in Mesozoic animal life, including dinosaurs.[84] Senter found that, contrary to popular depictions of roaring dinosaurs in motion pictures, it is likely that most dinosaurs were not capable of creating any vocalizations. To draw this conclusion, Senter studied the distribution of vocal organs in reptiles and birds. He found that vocal cords in the larynx probably evolved multiple times among reptiles, including crocodilians, which are able to produce guttural roars. Birds, on the other hand, lack a larynx. Instead, bird calls are produced by the syrinx, a vocal organ found only in birds, and which is not related to the larynx, meaning it evolved independently from the vocal organs in reptiles. The syrinx depends on the air sac system in birds to function; specifically, it requires the presence of a *clavicular air sac* near the wishbone or collar bone. This air sac leaves distinctive marks or opening on the bones, including a distinct opening in the upper arm bone (*humerus*). While many dinosaurs show evidence of extensive air sac systems, almost none possess the clavicular air sac necessary to vocalize (one exception, *Aerosteon*, probably evolved its clavicular air sac independently of birds for reasons other than vocalization).[84]

The most primitive animals with evidence of a vocalizing syrinx are the enantironithine birds. Any bird-line archosaurs more primitive than this probably did not make vocal calls. Rather, several lines of evidence suggest that dinosaurs used primarily visual communication, in the form of distinctive-looking (and possibly brightly colored) horns, frills, crests, sails and feathers. This is similar to some modern reptile groups such as lizards, in which many forms are largely silent (though like dinosaurs they possess well-developed senses of hearing) but use complex coloration and display behaviors to communicate.[84]

Also, though they may not have been able to vocalize, some dinosaurs may have used other methods of producing sound for communication. Modern animals, including reptiles and birds, use a wide variety of non-vocal sound communication, including hissing, jaw grinding or clapping, use of environment (such as splashing), and wing beating (which would have been possible in winged maniraptoran dinosaurs).[84]

Some studies have suggested that the hollow crests of the lambeosaurines may have functioned as resonance chambers used for a wide range of vocalizations.[85] [86] However, Senter (2008) noted that such chambers are also used in modern non-vocal animals to accentuate or deepen non-vocal sounds like hissing. For example, many snakes, which lack vocal cords, have resonating chambers in the skull.[84]

Physiology

Tyrannosaurus rex skull and upper vertebral column, Palais de la Découverte, Paris

A vigorous debate on the subject of temperature regulation in dinosaurs has been ongoing since the 1960s. Originally, scientists broadly disagreed as to whether dinosaurs were capable of regulating their body temperatures at all. More recently, dinosaur endothermy has become the consensus view, and debate has focused on the mechanisms of temperature regulation.

After dinosaurs were discovered, paleontologists first posited that they were ectothermic creatures: "terrible lizards" as their name suggests. This supposed cold-bloodedness was used to imply that dinosaurs were relatively slow, sluggish organisms, even though many modern reptiles are fast and light-footed despite relying on external sources of heat to regulate their body temperature. The idea of dinosaurs as ectothermic and sluggish remained a prevalent view until Robert T. "Bob" Bakker, an early proponent of dinosaur endothermy, published an influential paper on the topic in 1968.

Modern evidence indicates that dinosaurs thrived in cooler temperate climates, and that at least some dinosaur species must have regulated their body temperature by internal biological means (perhaps aided by the animals' bulk). Evidence of endothermy in dinosaurs includes the discovery of polar dinosaurs in Australia and Antarctica (where they would have experienced a cold, dark six-month winter), the discovery of dinosaurs whose feathers may have provided regulatory insulation, and analysis of blood-vessel structures within dinosaur bone that are typical of endotherms. Skeletal structures suggest that theropods and some other dinosaurs had active lifestyles better suited to an endothermic cardiovascular system, while sauropods exhibit fewer endothermic characteristics. It is certainly possible that some dinosaurs were endothermic while others were not. Scientific debate over the specifics continues.[87]

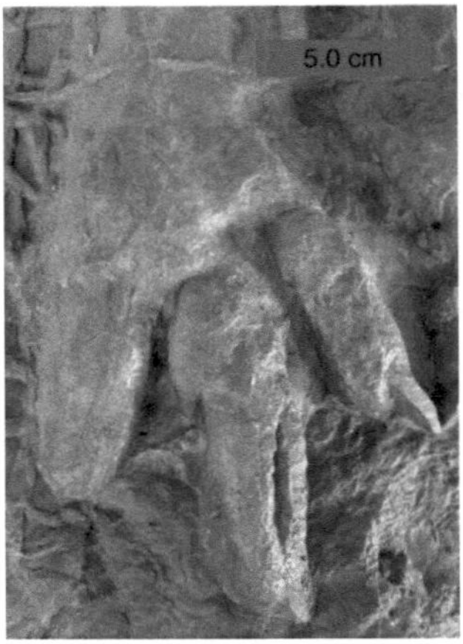

Eubrontes, a dinosaur footprint in the Lower Jurassic Moenave Formation at the St. George Dinosaur Discovery Site at Johnson Farm, southwestern Utah

Complicating the debate is the fact that warm-bloodedness can emerge based on more than one mechanism. Most discussions of dinosaur endothermy tend to compare them with average-sized birds or mammals, which expend energy to elevate body temperature above that of the environment. Small birds and mammals also possess insulation, such as fat, fur, or feathers, which slows down heat loss. However, large mammals, such as elephants, face a different problem because of their relatively small ratio of surface area to volume (Haldane's principle). This ratio compares the volume of an animal with the area of its skin: as an animal gets bigger, its surface area increases more slowly than its volume. At a certain point, the amount of heat radiated away through the skin drops below the amount of heat produced inside the body, forcing animals to use additional methods to avoid overheating. In the case of elephants, they have little hair as adults, have large ears which increase their surface area, and have behavioral adaptations as well (such as using the trunk to spray water on themselves and mud-wallowing). These behaviors increase cooling through evaporation.

Large dinosaurs would presumably have had to deal with similar issues; their body size suggest they lost heat relatively slowly to the surrounding air, and so could have been what are called inertial homeotherms, animals that are warmer than their environments through sheer size rather than through special adaptations like those of birds or mammals. However, so far this theory fails to account for the numerous dog- and goat-sized dinosaur species, or the young of larger species.

Modern computerized tomography (CT) scans of a dinosaur's chest cavity (conducted in 2000) found the apparent remnants of a four-chambered heart, much like those found in today's mammals and birds.[88] The idea is controversial within the scientific community, coming under fire for bad anatomical science[89] or simply wishful thinking.[90] The question of how this find reflects on metabolic rate and dinosaur internal anatomy may be moot, though, regardless of the object's identity: both modern crocodilians and birds, the closest living relatives of dinosaurs, have four-chambered hearts (albeit modified in crocodilians), and so dinosaurs probably had them as well.[91]

Soft tissue and DNA

One of the best examples of soft-tissue impressions in a fossil dinosaur was discovered in Petraroia, Italy. The discovery was reported in 1998, and described the specimen of a small, very young coelurosaur, *Scipionyx samniticus*. The fossil includes portions of the intestines, colon, liver, muscles, and windpipe of this immature dinosaur.[46]

In the March 2005 issue of *Science*, the paleontologist Mary Higby Schweitzer and her team announced the discovery of flexible material resembling actual soft tissue inside a 68-million-year-old *Tyrannosaurus rex* leg bone from the Hell Creek Formation in Montana. After recovery, the tissue was rehydrated by the science team.[47]

When the fossilized bone was treated over several weeks to remove mineral content from the fossilized bone-marrow cavity (a process called demineralization), Schweitzer found evidence of intact structures such as blood vessels, bone matrix, and connective tissue (bone fibers). Scrutiny under the microscope further revealed that the putative dinosaur soft tissue had retained fine structures (microstructures) even at the cellular level. The exact nature and composition of this material, and the implications of Schweitzer's discovery, are not yet clear; study and interpretation of the material is ongoing.[47]

Newer research, published in PloS One (30 July 2008), has challenged the claims that the material found is the soft tissue of *Tyrannosaurus*. Thomas Kaye of the University of Washington and his co-authors contend that what was really inside the tyrannosaur bone was slimy biofilm created by bacteria that coated the voids once occupied by blood vessels and cells.[92] The researchers found that what previously had been identified as remnants of blood cells, because of the presence of iron, were actually framboids, microscopic mineral spheres bearing iron. They found similar spheres in a variety of other fossils from various periods, including an ammonite. In the ammonite they found the spheres in a place where the iron they contain could not have had any relationship to the presence of blood.[93]

The successful extraction of ancient DNA from dinosaur fossils has been reported on two separate occasions, but, upon further inspection and peer review, neither of these reports could be confirmed.[94] However, a functional visual peptide of a theoretical dinosaur has been inferred using analytical phylogenetic reconstruction methods on gene sequences of related modern species such as reptiles and birds.[95] In addition, several proteins, including hemoglobin,[96] have putatively been detected in dinosaur fossils.[97]

Feathers and the origin of birds

The possibility that dinosaurs were the ancestors of birds was first suggested in 1868 by Thomas Henry Huxley.[98] After the work of Gerhard Heilmann in the early 20th century, the theory of birds as dinosaur descendants was abandoned in favor of the idea of their being descendants of generalized thecodonts, with the key piece of evidence being the supposed lack of clavicles in dinosaurs.[99] However, as later discoveries showed, clavicles (or a single fused wishbone, which derived from separate clavicles) were not actually absent;[14] they had been found as early as 1924 in *Oviraptor*, but misidentified as an interclavicle.[100] In the 1970s, John Ostrom revived the dinosaur–bird theory,[101] which gained momentum in the coming decades with the advent of cladistic analysis,[102] and a great increase in the discovery of small theropods and early birds.[22] Of particular note have been the fossils of the Yixian Formation, where a variety of theropods and early birds have been found, often with feathers of some type.[14] Birds share over a hundred distinct anatomical features with theropod dinosaurs, which are now generally accepted to have been their closest ancient relatives.[103] They are most closely allied with maniraptoran coelurosaurs.[14] A minority of scientists, most notably Alan Feduccia and Larry Martin, have proposed other evolutionary paths, including revised versions of Heilmann's basal archosaur proposal,[104] or that maniraptoran theropods are the ancestors of birds but themselves are not dinosaurs, only convergent with dinosaurs.[105]

Feathers

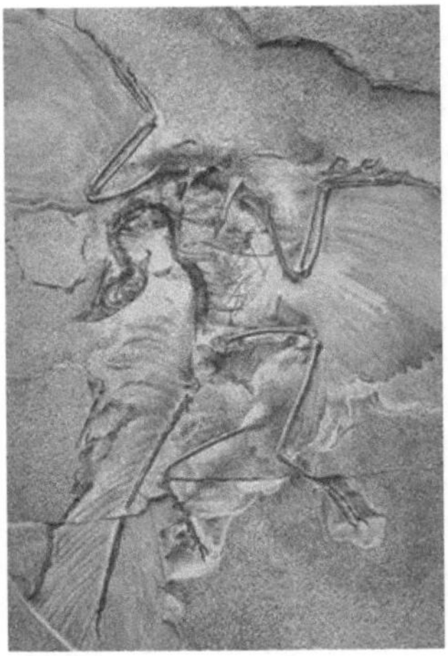

The famous Berlin Specimen of *Archaeopteryx lithographica*

Archaeopteryx, the first good example of a "feathered dinosaur", was discovered in 1861. The initial specimen was found in the Solnhofen limestone in southern Germany, which is a *lagerstätte*, a rare and remarkable geological formation known for its superbly detailed fossils. *Archaeopteryx* is a transitional fossil, with features clearly intermediate between those of modern reptiles and birds. Brought to light just two years after Darwin's seminal *The Origin of Species*, its discovery spurred the nascent debate between proponents of evolutionary biology and creationism. This early bird is so dinosaur-like that, without a clear impression of feathers in the surrounding rock, at least one specimen was mistaken for *Compsognathus*.[106]

Since the 1990s, a number of additional feathered dinosaurs have been found, providing even stronger evidence of the close relationship between dinosaurs and modern birds. Most of these specimens were unearthed in the lagerstätte of the Yixian Formation, Liaoning, northeastern China, which was part of an island continent during the Cretaceous. Though feathers have been found in only a few locations, it is possible that non-avian dinosaurs elsewhere in the world were also feathered. The lack of widespread fossil evidence for feathered non-avian dinosaurs may be because delicate features like skin and feathers are not often preserved by fossilization and thus are absent from the fossil record. To this point, protofeathers (thin, filament-like structures) are known from dinosaurs at the base of Coelurosauria, such as compsognathids like *Sinosauropteryx* and tyrannosauroids (*Dilong*),[107] but barbed feathers are known only among the coelurosaur subgroup Maniraptora, which includes oviraptorosaurs, troodontids, dromaeosaurids, and birds.[14] [108] The description of feathered dinosaurs has not been without controversy; perhaps the most vocal critics have been Alan Feduccia and Theagarten Lingham-Soliar, who have proposed that protofeathers are the result of the decomposition of collagenous fiber that underlaid the dinosaurs' integument,[109] [110] [111] and that maniraptoran dinosaurs with barbed feathers were not actually dinosaurs, but convergent with dinosaurs.[105] [110] However, their views have for the most part not been accepted by other researchers, to the point that the question of the scientific nature of Feduccia's proposals has been raised.[112]

Skeleton

Because feathers are often associated with birds, feathered dinosaurs are often touted as the missing link between birds and dinosaurs. However, the multiple skeletal features also shared by the two groups represent another important line of evidence for paleontologists. Areas of the skeleton with important similarities include the neck, pubis, wrist (semi-lunate carpal), arm and pectoral girdle, furcula (wishbone), and breast bone. Comparison of bird and dinosaur skeletons through cladistic analysis strengthens the case for the link.

Soft anatomy

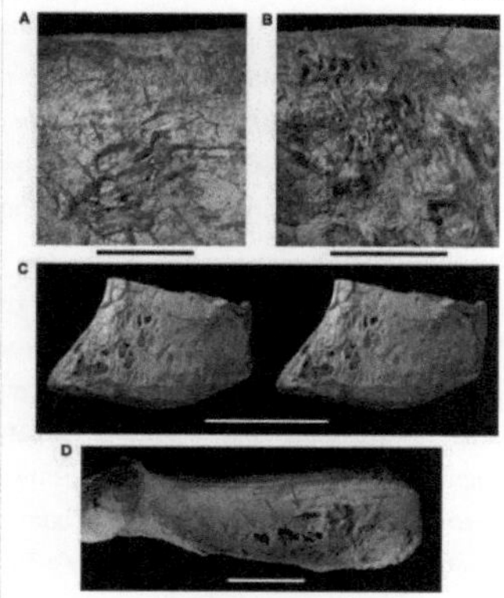

Pneumatopores on the left ilium of *Aerosteon riocoloradensis*

Large meat-eating dinosaurs had a complex system of air sacs similar to those found in modern birds, according to an investigation which was led by Patrick O'Connor of Ohio University. The lungs of theropod dinosaurs (carnivores that walked on two legs and had bird-like feet) likely pumped air into hollow sacs in their skeletons, as is the case in birds. "What was once formally considered unique to birds was present in some form in the ancestors of birds", O'Connor said.[113] In a 2008 paper published in the online journal *PLoS ONE*, scientists described *Aerosteon riocoloradensis*, the skeleton of which supplies the strongest evidence to date of a dinosaur with a bird-like breathing system. CT-scanning revealed the evidence of air sacs within the body cavity of the *Aerosteon* skeleton.[114] [115]

Reproductive biology

A discovery of features in a *Tyrannosaurus rex* skeleton provided evidence of medullary bone in dinosaurs and, for the first time, allowed paleontologists to establish the sex of a dinosaur. When laying eggs, female birds grow a special type of bone between the hard outer bone and the marrow of their limbs. This medullary bone, which is rich in calcium, is used to make eggshells. The presence of endosteally derived bone tissues lining the interior marrow cavities of portions of the *Tyrannosaurus rex* specimen's hind limb suggested that *T. rex* used similar reproductive strategies, and revealed the specimen to be female.[116] Further research has found medullary bone in the theropod *Allosaurus* and the ornithopod *Tenontosaurus*. Because the line of dinosaurs that includes *Allosaurus* and *Tyrannosaurus* diverged from the line that led to *Tenontosaurus* very early in the evolution of dinosaurs, this suggests that dinosaurs in general produced medullary tissue. Medullary bone has been found in specimens of sub-adult size, which suggests that dinosaurs reached sexual maturity rather quickly for such large animals.[117]

Behavioral evidence

Fossils of the troodonts *Mei* and *Sinornithoides* demonstrate that some dinosaurs slept with their heads tucked under their arms.[118] This behavior, which may have helped to keep the head warm, is also characteristic of modern birds. Several deinonychosaur and oviraptorosaur specimens have also been found preserved on top of their nests, likely brooding in a bird-like manner.[119] The ratio between egg volume and body mass of adults among these dinosaurs suggest that the eggs were primarily brooded by the male, and that the young were highly precocial, similar to many modern ground-dwelling birds.[120]

Some dinosaurs are known to have used gizzard stones like modern birds. These stones are swallowed by animals to aid digestion and break down food and hard fibers once they enter the stomach. When found in association with fossils, gizzard stones are called gastroliths.[121]

Extinction of major groups

The discovery that birds are a type of dinosaur showed that dinosaurs in general are not, in fact, extinct as is commonly stated.[122] However, all non-avian dinosaurs as well as many groups of birds did suddenly become extinct approximately 65 million years ago. Many other groups of animals also became extinct at this time, including ammonites (nautilus-like mollusks), mosasaurs, plesiosaurs, pterosaurs, and many groups of mammals.[6] This mass extinction is known as the Cretaceous–Paleogene extinction event. The nature of the event that caused this mass extinction has been extensively studied since the 1970s; at present, several related theories are supported by paleontologists. Though the consensus is that an impact event was the primary cause of dinosaur extinction, some scientists cite other possible causes, or support the idea that a confluence of several factors was responsible for the sudden disappearance of dinosaurs from the fossil record.

At the peak of the Mesozoic, there were no polar ice caps, and sea levels are estimated to have been from 100 to 250 meters (300 to 800 ft) higher than they are today. The planet's temperature was also much more uniform, with only 25 °C (45 °F) separating average polar temperatures from those at the equator. On average, atmospheric temperatures were also much higher; the poles, for example, were 50 °C (90 °F) warmer than today.[123] [124]

The atmosphere's composition during the Mesozoic was vastly different as well. Carbon dioxide levels were up to 12 times higher than today's levels, and oxygen formed 32 to 35% of the atmosphere, as compared to 21% today. However, by the late Cretaceous, the environment was changing dramatically. Volcanic activity was decreasing, which led to a cooling trend as levels of atmospheric carbon dioxide dropped. Oxygen levels in the atmosphere also started to fluctuate and would ultimately fall considerably. Some scientists hypothesize that climate change, combined with lower oxygen levels, might have led directly to the demise of many species. If the dinosaurs had respiratory systems similar to those commonly found in modern birds, it may have been particularly difficult for them to cope with reduced respiratory efficiency, given the enormous oxygen demands of their very large bodies.[6]

Impact event

The Chicxulub Crater at the tip of the Yucatán Peninsula; the impactor that formed this crater may have caused the dinosaur extinction.

The asteroid collision theory, which was brought to wide attention in 1980 by Walter Alvarez and colleagues, links the extinction event at the end of the Cretaceous period to a bolide impact approximately 65.5 million years ago. Alvarez *et al.* proposed that a sudden increase in iridium levels, recorded around the world in the period's rock stratum, was direct evidence of the impact.[125] The bulk of the evidence now suggests that a bolide 5 to 15 kilometers (3 to 9 mi) wide hit in the vicinity of the Yucatán Peninsula (in southeastern Mexico), creating the approximately 180 km (110 mi) Chicxulub Crater and triggering the mass extinction.[126] [127] Scientists are not certain whether dinosaurs were thriving or declining before the impact event. Some scientists propose that the meteorite caused a long and unnatural drop in Earth s atmospheric temperature, while others claim that it would have instead created an unusual heat wave. The consensus among scientists who support this theory is that the impact caused extinctions both directly (by heat from the meteorite impact) and also indirectly (via a worldwide cooling brought about when matter ejected from the impact crater reflected thermal radiation from the sun). Although the speed of extinction cannot be deduced from the fossil record alone, various models suggest that the extinction was extremely rapid, being down to hours rather than years.[128]

In September 2007, U.S. researchers led by William Bottke of the Southwest Research Institute in Boulder, Colorado, and Czech scientists used computer simulations to identify the probable source of the Chicxulub impact. They calculated a 90% probability that a giant asteroid named Baptistina, approximately 160 km (99 mi) in diameter, orbiting in the asteroid belt which lies between Mars and Jupiter, was struck by a smaller unnamed asteroid about 55 km (35 mi) in diameter about 160 million years ago. The impact shattered Baptistina, creating a cluster which still exists today as the Baptistina family. Calculations indicate that some of the fragments were sent hurtling into earth-crossing orbits, one of which was the 10 km (6.2 mi) wide meteorite which struck Mexico's Yucatan peninsula 65 million years ago, creating the Chicxulub crater.[129] In 2011, new data from the Wide-field Infrared Survey Explorer revised the date of the collision which created the Baptistina family to about 80 million years ago. This makes an asteroid from this family highly improbable to be the asteroid that created the Chicxulub Crater, as typically the process of resonance and collision of an asteroid takes many tens of millions of years.[130]

A similar but more controversial explanation proposes that "passages of the [hypothetical] solar companion star Nemesis through the Oort comet cloud would trigger comet showers."[131] One or more of these comets then collided with the Earth at approximately the same time, causing the worldwide extinction. As with the impact of a single asteroid, the end result of this comet bombardment would have been a sudden drop in global temperatures, followed by a protracted cool period.[131]

Deccan Traps

Before 2000, arguments that the Deccan Traps flood basalts caused the extinction were usually linked to the view that the extinction was gradual, as the flood basalt events were thought to have started around 68 million years ago and lasted for over 2 million years. However, there is evidence that two thirds of the Deccan Traps were created in only 1 million years about 65.5 million years ago, and so these eruptions would have caused a fairly rapid extinction, possibly over a period of thousands of years, but still longer than would be expected from a single impact event.[132][133]

The Deccan Traps could have caused extinction through several mechanisms, including the release into the air of dust and sulphuric aerosols, which might have blocked sunlight and thereby reduced photosynthesis in plants. In addition, Deccan Trap volcanism might have resulted in carbon dioxide emissions, which would have increased the greenhouse effect when the dust and aerosols cleared from the atmosphere.[133] Before the mass extinction of the dinosaurs, the release of volcanic gases during the formation of the Deccan Traps "contributed to an apparently massive global warming. Some data point to an average rise in temperature of 8 °C (14 °F) in the last half million years before the impact [at Chicxulub]."[132] [133]

In the years when the Deccan Traps theory was linked to a slower extinction, Luis Alvarez (who died in 1988) replied that paleontologists were being misled by sparse data. While his assertion was not initially well-received, later intensive field studies of fossil beds lent weight to his claim. Eventually, most paleontologists began to accept the idea that the mass extinctions at the end of the Cretaceous were largely or at least partly due to a massive Earth impact. However, even Walter Alvarez has acknowledged that there were other major changes on Earth even before the impact, such as a drop in sea level and massive volcanic eruptions that produced the Indian Deccan Traps, and these may have contributed to the extinctions.[134]

Failure to adapt to changing conditions

Lloyd *et al.* (2008) noted that, in the Mid Cretaceous, the flowering, angiosperm plants became a major part of terrestrial ecosystems, which had previously been dominated by gymnosperms such as conifers. Dinosaur coprolite–fossilized dung–indicate that, while some ate angiosperms, most herbivorous dinosaurs ate mainly gymnosperms. Statistical analysis by Lloyd *et al.* concluded that, contrary to earlier studies, dinosaurs did not diversify very much in the Late Cretaceous. Lloyd *et al.* suggested that dinosaurs' failure to diversify as ecosystems were changing doomed them to extinction.[135]

Possible Paleocene survivors

Non-avian dinosaur remains are occasionally found above the K–T boundary. In 2001, paleontologists Zielinski and Budahn reported the discovery of a single hadrosaur leg-bone fossil in the San Juan Basin, New Mexico, and described it as evidence of Paleocene dinosaurs. The formation in which the bone was discovered has been dated to the early Paleocene epoch, approximately 64.5 million years ago. If the bone was not re-deposited into that stratum by weathering action, it would provide evidence that some dinosaur populations may have survived at least a half million years into the Cenozoic Era.[136] Other evidence includes the finding of dinosaur remains in the Hell Creek Formation up to 1.3 meters (51 in) above (40000 years later than) the K–T boundary. Similar reports have come from other parts of the world, including China.[137] Many scientists, however, dismissed the supposed Paleocene dinosaurs as re-worked, that is, washed out of their original locations and then re-buried in much later sediments.[138] [139] However, direct dating of the bones themselves has supported the later date, with U–Pb dating methods resulting in a precise age of 64.8 ± 0.9 million years ago.[140] If correct, the presence of a handful of dinosaurs in the early Paleocene would not change the underlying facts of the extinction.[138]

History of discovery

Dinosaur fossils have been known for millennia, although their true nature was not recognized. The Chinese, whose modern word for dinosaur is *konglong* (, or "terrible dragon"), considered them to be dragon bones and documented them as such. For example, *Hua Yang Guo Zhi*, a book written by Zhang Qu during the Western Jin Dynasty, reported the discovery of dragon bones at Wucheng in Sichuan Province.[141] Villagers in central China have long unearthed fossilized "dragon bones" for use in traditional medicines, a practice that continues today.[142] In Europe, dinosaur fossils were generally believed to be the remains of giants and other creatures killed by the Great Flood.

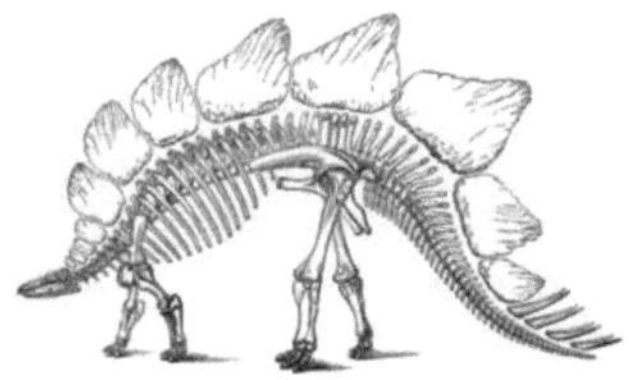

Marsh's 1896 illustration of the bones of *Stegosaurus*, a dinosaur he described and named in 1877.

Scholarly descriptions of what would now be recognized as dinosaur bones first appeared in the late 17th century in England. Part of a bone, now known to have been the femur of a *Megalosaurus*,[143] was recovered from a limestone quarry at Cornwell near Chipping Norton, Oxfordshire, England, in 1676. The fragment was sent to Robert Plot, Professor of Chemistry at the University of Oxford and first curator of the Ashmolean Museum, who published a description in his *Natural History of Oxfordshire* in 1677. He correctly identified the bone as the lower extremity of the femur of a large animal, and recognized that it was too large to belong to any known species. He therefore concluded it to be the thigh bone of a giant human similar to those mentioned in the Bible. In 1699, Edward Lhuyd, a friend of Sir Isaac Newton, was responsible for the first published scientific treatment of what would now be recognized as a dinosaur when he described and named a sauropod tooth, "Rutellum implicatum",[144] [145] that had been found in Caswell, near Witney, Oxfordshire.[146]

William Buckland

Between 1815 and 1824, the Rev William Buckland, a professor of geology at Oxford University, collected more fossilized bones of *Megalosaurus* and became the first person to describe a dinosaur in a scientific journal.[143] [147] The second dinosaur genus to be identified, *Iguanodon*, was discovered in 1822 by Mary Ann Mantell – the wife of English geologist Gideon Mantell. Gideon Mantell recognized similarities between his fossils [148] and the bones of modern iguanas. He published his findings in 1825.[149] [150]

The study of these "great fossil lizards" soon became of great interest to European and American scientists, and in 1842 the English paleontologist Richard Owen coined the term "dinosaur". He recognized that the remains that had been found so far, *Iguanodon*, *Megalosaurus* and *Hylaeosaurus*, shared a number of distinctive features, and so decided to present them as a distinct taxonomic group. With the backing of Prince Albert of Saxe-Coburg-Gotha, the husband of Queen Victoria, Owen established the Natural History Museum in South Kensington, London, to display the national collection of dinosaur fossils and other biological and geological exhibits.

In 1858, the first known American dinosaur was discovered, in marl pits in the small town of Haddonfield, New Jersey (although fossils had been found before, their nature had not been correctly discerned). The creature was named *Hadrosaurus foulkii*. It was an extremely important find: *Hadrosaurus* was one of the first nearly complete dinosaur skeletons found (the first was in 1834, in Maidstone, Kent, England), and it was clearly a bipedal creature. This was a revolutionary discovery as, until that point, most scientists had believed dinosaurs walked on four feet, like other lizards. Foulke's discoveries sparked a wave of dinosaur mania in the United States.

Othniel Charles Marsh, 19th century photograph

Dinosaur mania was exemplified by the fierce rivalry between Edward Drinker Cope and Othniel Charles Marsh, both of whom raced to be the first to find new dinosaurs in what came to be known as the Bone Wars. The feud probably originated when Marsh publicly pointed out that Cope's reconstruction of an *Elasmosaurus* skeleton was flawed: Cope had inadvertently placed the plesiosaur's head at what should have been the animal's tail end. The fight between the two scientists lasted for over 30 years, ending in 1897 when Cope died after spending his entire fortune on the dinosaur hunt. Marsh 'won' the contest primarily because he was better funded through a relationship with the US Geological Survey. Unfortunately, many valuable dinosaur specimens were damaged or destroyed due to the pair's rough methods: for example, their diggers often used dynamite to unearth bones (a method modern paleontologists would find appalling). Despite their unrefined methods, the contributions of Cope and Marsh to paleontology were vast: Marsh unearthed 86 new species of dinosaur and Cope discovered 56, a total of 142 new species. Cope's collection is now at the American Museum of Natural History in New York, while Marsh's is on display at the Peabody Museum of Natural History at Yale University.[151]

Edward Drinker Cope, 19th century photograph

After 1897, the search for dinosaur fossils extended to every continent, including Antarctica. The first Antarctic dinosaur to be discovered, the ankylosaurid *Antarctopelta oliveroi*, was found on Ross Island in 1986, although it was 1994 before an Antarctic species, the theropod *Cryolophosaurus ellioti*, was formally named and described in a scientific journal.

Current dinosaur "hot spots" include southern South America (especially Argentina) and China. China in particular has produced many exceptional feathered dinosaur specimens due to the unique geology of its dinosaur beds, as well as an ancient arid climate particularly conducive to fossilization.

The "dinosaur renaissance"

The field of dinosaur research has enjoyed a surge in activity that began in the 1970s and is ongoing. This was triggered, in part, by John Ostrom's discovery of *Deinonychus*, an active predator that may have been warm-blooded, in marked contrast to the then-prevailing image of dinosaurs as sluggish and cold-blooded. Vertebrate paleontology has become a global science. Major new dinosaur discoveries have been made by paleontologists working in previously unexploited regions, including India, South America, Madagascar, Antarctica, and most significantly China (the amazingly well-preserved feathered dinosaurs in China have further consolidated the link between dinosaurs and their conjectured living descendants, modern birds). The widespread application of cladistics, which rigorously analyzes the relationships between biological organisms, has also proved tremendously useful in classifying dinosaurs. Cladistic analysis, among other modern techniques, helps to compensate for an often incomplete and fragmentary fossil record.

Cultural depictions

By human standards, dinosaurs were creatures of fantastic appearance and often enormous size. As such, they have captured the popular imagination and become an enduring part of human culture. Entry of the word "dinosaur" into the common vernacular reflects the animals' cultural importance: in English, "dinosaur" is commonly used to describe anything that is impractically large, obsolete, or bound for extinction.[7]

Public enthusiasm for dinosaurs first developed in Victorian England, where in 1854, three decades after the first scientific descriptions of dinosaur remains, the famous dinosaur sculptures were unveiled in London's Crystal Palace Park. The Crystal Palace dinosaurs proved so popular that a strong market in smaller replicas soon developed. In subsequent decades, dinosaur exhibits opened at parks and museums around the world, ensuring that successive generations would be introduced to the animals in an immersive and exciting way.[152] Dinosaurs' enduring popularity, in its turn, has resulted in significant public funding for dinosaur science, and has frequently spurred new discoveries. In the United States, for example, the competition between museums for public attention led directly to the Bone Wars of the 1880s and 1890s, during which a pair of feuding paleontologists made enormous scientific contributions.[153]

The popular preoccupation with dinosaurs has ensured their appearance in literature, film and other media. Beginning in 1852 with a passing mention in Charles Dickens' *Bleak House*,[154] dinosaurs have been featured in large numbers of fictional works. Sir Arthur Conan Doyle's 1912 book *The Lost World*, the iconic 1933 film *King Kong*, 1954's *Godzilla* and its many sequels, the best-selling 1990 novel *Jurassic Park* by Michael Crichton and its 1993 film adaptation are just a few notable examples of dinosaur appearances in fiction. Authors of general-interest non-fiction works about dinosaurs, including some prominent paleontologists, have often sought to use the animals as a way to educate readers about science in general. Dinosaurs are ubiquitous in advertising; numerous companies have referenced dinosaurs in printed or televised advertisements, either in order to sell their own products or in order to characterize their rivals as slow-moving, dim-witted or obsolete.[155]

Notes and references

[1] Gauthier, Jacques; de Querioz, Kevin (2001). "Feathered dinosaurs, flying dinosaurs, crown dinosaurs, and the name 'Aves'." (http://vertebrates.si.edu/herps/herps_pdfs/deQueiroz_pdfs/2001gaudeqost.pdf) (PDF). *New Perspectives on the Origin and Early Evolution of Birds: Proceedings of the International Symposium in Honor of John H. Ostrom*. Peabody Museum of Natural History, Yale University. ISBN 0-912532-57-2. . Retrieved 2009-09-22.

[2] Zhou, Z. (2004). "The origin and early evolution of birds: discoveries, disputes, and perspectives from fossil evidence". *Naturwissenchaften* **91** (10): 455–471. Bibcode 2004NW.....91..455Z. doi:10.1007/s00114-004-0570-4.

[3] Alfaro, M.E., F. Santini, C. Brock, H. Alamillo, A. Dornburg. D.L. Rabosky, G. Carnevale, and L.J. Harmon (2009). "Nine exceptional radiations plus high turnover explain species diversity in jawed vertebrates". *Proceedings of the National Academy of Sciences USA* **106** (32): 13410–13414. Bibcode 2009PNAS..10613410A. doi:10.1073/pnas.0811087106. PMC 2715324. PMID 19633192.

[4] Wang, S.C., and Dodson, P. (2006). "Estimating the Diversity of Dinosaurs". *Proceedings of the National Academy of Sciences USA* **103** (37): 13601–13605. Bibcode 2006PNAS..10313601W. doi:10.1073/pnas.0606028103. PMC 1564218. PMID 16954187.

[5] Amos J (2008-09-17). "Will the real dinosaurs stand up?" (http://news.bbc.co.uk/2/hi/science/nature/7620621.stm). *BBC News*. . Retrieved 2011-03-23.

[6] MacLeod, N, Rawson, PF, Forey, PL, Banner, FT, Boudagher-Fadel, MK, Bown, PR, Burnett, JA, Chambers, P, Culver, S, Evans, SE, Jeffery, C, Kaminski, MA, Lord, AR, Milner, AC, Milner, AR, Morris, N, Owen, E, Rosen, BR, Smith, AB, Taylor, PD, Urquhart, E & Young, JR (1997). "The Cretaceous–Tertiary biotic transition" (http://findarticles.com/p/articles/mi_qa3721/is_199703/ai_n8738406/print). *Journal of the Geological Society* **154** (2): 265–292. doi:10.1144/gsjgs.154.2.0265. .

[7] "Dinosaur – Definition and More" (http://www.m-w.com/dictionary/dinosaur). Merriam-Webster Dictionary. . Retrieved 2011-05-06.

[8] Owen, R (1842)). *Report on British Fossil Reptiles." Part II.* Report of the Eleventh Meeting of the British Association for the Advancement of Science; Held at Plymouth in July 1841 (http://books.google.com/books?id=dy5LAAAAYAAJ&pg=PA60&f=false#v=onepage&q&f=false). London: John Murray. pp. 60–204. .

[9] "Liddell–Scott–Jones Lexicon of Classical Greek" (http://www.perseus.tufts.edu/cgi-bin/lexindex?lookup=deino/s&lang=greek&doc=Perseus:text:1999.01.0169&formentry=0). . Retrieved 2008-08-05.

[10] Farlow, J.O., and Brett-Surman, M.K. (1997). "Preface". In Farlow, J.O., and Brett-Surman, M.K. (eds.). *The Complete Dinosaur*. Indiana University Press. pp. ix–xi. ISBN 0-253-33349-0.

[11] Benton, Michael J. (2004). "Origin and relationships of Dinosauria". In Weishampel, David B.; Dodson, Peter; and Osmólska, Halszka (eds.). *The Dinosauria* (2nd ed.). Berkeley: University of California Press. pp. 7–19. ISBN 0-520-24209-2.

[12] Olshevsky, G. (2000). "An annotated checklist of dinosaur species by continent". *Mesozoic Meanderings* **3**: 1–157.

[13] Sereno, P. (2005). "The logical basis of phylogenetic taxonomy". *Systematic Biology* **54** (4): 595–619.

[14] Padian K (2004). "Basal avialae". In Weishampel DB, Dodson P, Osmólska H. *The Dinosauria (2d edition)*. University of California Press. pp. 210–231. ISBN 0-520-24209-2.

[15] Glut, Donald F. (1997). *Dinosaurs: The Encyclopedia*. Jefferson, North Carolina: McFarland & Co. p. 40. ISBN 0-89950-917-7.

[16] Lambert, David; and the Diagram Group (1990). *The Dinosaur Data Book*. New York: Avon Books. p. 288. ISBN 0-380-75896-2.

[17] Morales, Michael (1997). "Nondinosaurian vertebrates of the Mesozoic". In Farlow JO, Brett-Surman MK. *The Complete Dinosaur*. Bloomington: Indiana University Press. pp. 607–624. ISBN 0-253-33349-0.

[18] Hu Yaoming; Meng, J; Wang, Y; Li, C (2005). "Large Mesozoic mammals fed on dinosaurs". *Nature* **433** (7022): 149–152. Bibcode 2005Natur.433..149H. doi:10.1038/nature03102. PMID 15650737.

[19] Russell, Dale A. (1995). "China and the lost worlds of the dinosaurian era". *Historical Biology* **10**: 3–12. doi:10.1080/10292389509380510.

[20] Amiot, R.; Buffetaut, E.; Lécuyer, C.; Wang, X.; Boudad, L.; Ding, Z.; Fourel, F.; Hutt, S.; Martineau, F.; Medeiros, A.; Mo, J.; Simon, L.; Suteethorn, V.; Sweetman, S.; Tong, H.; Zhang, F.; and Zhou, Z. (2010). "Oxygen isotope evidence for semi-aquatic habits among spinosaurid theropods". *Geology* **38** (2): 139–142. doi:10.1130/G30402.1.

[21] Nesbitt S.J. (2011). "The early evolution of archosaurs : relationships and the origin of major clades". *Bulletin of the American Museum of Natural History* **352**: 1–292. doi:10.1206/352.1.

[22] Holtz, Jr., T.R. (2000). "Classification and evolution of the dinosaur groups". In Paul, G.S.. *The Scientific American Book of Dinosaurs*. St. Martin's Press. pp. 140–168. ISBN 0-312-26226-4.

[23] Langer, M.C., Abdala, F., Richter, M., and Benton, M.J. (1999). "A sauropodomorph dinosaur from the Upper Triassic (Carnian) of southern Brazil". *Comptes Rendus de l'Academie des Sciences, Paris: Sciences de la terre et des planètes* **329**: 511–517.

[24] Nesbitt, Sterling J.; Irmis, Randall B.; Parker, William G. (2007). "A critical re-evaluation of the Late Triassic dinosaur taxa of North America". *Journal of Systematic Palaeontology* **5** (2): 209–243. doi:10.1017/S1477201907002040.

[25] This was recognized not later than 1909: "Dr. Holland and the Sprawling Sauropods" (http://www.hmnh.org/library/diplodocus/holland1910.html). . The arguments and many of the images are also presented in Desmond, A. (1976). *Hot Blooded Dinosaurs*. DoubleDay. ISBN 0-385-27063-1.

[26] Benton, M.J. (2004). *Vertebrate Paleontology*. Blackwell Publishers. xii–452. ISBN 0-632-05614-2.

[27] Cowen, Richard (2004). "Dinosaurs". *History of Life* (4th ed.). Blackwell Publishing. pp. 151–175. ISBN 1-4051-1756-7. OCLC 53970577.

[28] Kubo, T.; Benton, Michael J. (2007). "Evolution of hindlimb posture in archosaurs: limb stresses in extinct vertebrates". *Palaeontology* **50** (6): 1519–1529. doi:10.1111/j.1475-4983.2007.00723.x.

[29] Seeley, H.G. (1887). "On the classification of the fossil animals commonly named Dinosauria". *Proc R Soc London* (Royal Society) **43**: 165–171. Bibcode 1887RSPS...43..165S. doi:10.1098/rspl.1887.0117.

[30] Romer, A.S. (1956). *Osteology of the Reptiles*. University of Chicago. ISBN 0-89464-985-X.

[31] Ostrom, J.H. (1980). "The evidence of endothermy in dinosaurs". In Thomas, R.D.K. and Olson, E.C.. *A cold look at the warm-blooded dinosaurs*. Boulder, CO: American Association for the Advancement of Science. pp. 82–105.

[32] Bakker, R. T., and Galton, P (1974). "Dinosaur monophyly and a new class of vertebrates". *Nature* **248** (5444): 168–172. Bibcode 1974Natur.248..168B. doi:10.1038/248168a0.

[33] Kump LR, Pavlov A & Arthur MA (2005). "Massive release of hydrogen sulfide to the surface ocean and atmosphere during intervals of oceanic anoxia". *Geology* **33** (5): 397–400. Bibcode 2005Geo....33..397K. doi:10.1130/G21295.1.

[34] Tanner LH, Lucas SG & Chapman MG (2004). "Assessing the record and causes of Late Triassic extinctions" (http://web.archive.org/web/20071025225841/http://nmnaturalhistory.org/pdf_files/TJB.pdf) (PDF). *Earth-Science Reviews* **65** (1–2): 103–139. Bibcode 2004ESRv...65..103T. doi:10.1016/S0012-8252(03)00082-5. Archived from the original (http://nmnaturalhistory.org/pdf_files/TJB.pdf) on October 25, 2007. . Retrieved 2007-10-22.

[35] Sereno PC (1999). "The evolution of dinosaurs". *Science* **284** (5423): 2137–2147. doi:10.1126/science.284.5423.2137. PMID 10381873.

[36] Sereno, P.C.; Forster, Catherine A.; Rogers, Raymond R.; Monetta, Alfredo M. (1993). "Primitive dinosaur skeleton from Argentina and the early evolution of Dinosauria". *Nature* **361** (6407): 64–66. Bibcode 1993Natur.361...64S. doi:10.1038/361064a0.

[37] Brusatte, S.L.; Benton, MJ; Ruta, M; Lloyd, GT (2008). "Superiority, competition, and opportunism in the evolutionary radiation of dinosaurs". *Science* **321** (5895): 1485–1488. Bibcode 2008Sci...321.1485B. doi:10.1126/science.1161833. PMID 18787166.

[38] Paul, G.S. (1988). *Predatory Dinosaurs of the World*. New York: Simon and Schuster. pp. 248–250. ISBN 0-671-61946-2.

[39] Clark J.M., Maryanska T., Barsbold R (2004). "Therizinosauroidea". In Weishampel DB, Dodson P, Osmólska H. *The Dinosauria (2d edition)*. University of California Press. pp. 151–164. ISBN 0-520-24209-2.

[40] Norell MA, Makovicky PJ (2004). "Dromaeosauridae". In Weishampel DB, Dodson P, Osmólska H. *The Dinosauria (2d edition)*. University of California Press. pp. 196–210. ISBN 0-520-24209-2.

[41] Holtz, Thomas R., Jr.; Chapman, Ralph E.; and Lamanna, Matthew C. (2004). "Mesozoic biogeography of Dinosauria". In Weishampel, David B.; Dodson, Peter; and Osmólska, Halszka (eds.). *The Dinosauria* (2nd ed.). Berkeley: University of California Press. pp. 627–642. ISBN 0-520-24209-2.

[42] Fastovsky, David E.; and Smith, Joshua B. (2004). "Dinosaur paleoecology". In Weishampel, David B.; Dodson, Peter; and Osmólska, Halszka. *The Dinosauria* (2nd ed.). Berkeley: University of California Press. pp. 614–626. ISBN 0-520-24209-2.

[43] Sereno, P.C.; Wilson, JA; Witmer, LM; Whitlock, JA; Maga, A; Ide, O; Rowe, TA; Kemp, Tom (2007). "Structural extremes in a Cretaceous dinosaur". *PLoS ONE* **2** (11): e1230. Bibcode 2007PLoSO...2.1230S. doi:10.1371/journal.pone.0001230. PMC 2077925. PMID 18030355.

[44] Prasad, V.; Strömberg, CA; Alimohammadian, H; Sahni, A (2005). "Dinosaur coprolites and the early evolution of grasses and grazers". *Science* **310** (5751): 1170–1180. Bibcode 2005Sci...310.1177P. doi:10.1126/science.1118806. PMID 16293759.

[45] Archibald, J. David; and Fastovsky, David E. (2004). "Dinosaur Extinction". In Weishampel, David B.; Dodson, Peter; and Osmólska, Halszka (eds.). *The Dinosauria* (2nd ed.). Berkeley: University of California Press. pp. 672–684. ISBN 0-520-24209-2.

[46] Dal Sasso, C. and Signore, M. (1998). "Exceptional soft-tissue preservation in a theropod dinosaur from Italy". *Nature* **292** (6674): 383–387. Bibcode 1998Natur.392..383D. doi:10.1038/32884.

[47] Schweitzer, M.H., Wittmeyer, J.L. and Horner, J.R. (2005). "Soft-Tissue Vessels and Cellular Preservation in Tyrannosaurus rex". *Science* **307** (5717): 1952–1955. Bibcode 2005Sci...307.1952S. doi:10.1126/science.1108397. PMID 15790853.

[48] Carpenter, Kenneth (2006). "Biggest of the big: a critical re-evaluation of the mega-sauropod *Amphicoelias fragillimus*". In Foster, John R.; and Lucas, Spencer G. (eds.). *Paleontology and Geology of the Upper Jurassic Morrison Formation*. New Mexico Museum of Natural History and Science Bulletin **36**. Albuquerque: New Mexico Museum of Natural History and Science. pp. 131–138.

[49] Farlow JA (1993). "On the rareness of big, fierce animals: speculations about the body sizes, population densities, and geographic ranges of predatory mammals and large, carnivorous dinosaurs". In Dodson, Peter; and Gingerich, Philip. *Functional Morphology and Evolution*. American Journal of Science, Special Volume **293-A**. pp. 167–199.

[50] Peczkis, J. (1994). "Implications of body-mass estimates for dinosaurs". *Journal of Vertebrate Paleontology* **14** (4): 520–33. doi:10.1080/02724634.1995.10011575.

[51] "Anatomy and evolution" (http://paleobiology.si.edu/dinosaurs/info/everything/evo_1.html). National Museum of Natural History. . Retrieved 2007-11-21.

[52] Colbert, Edwin Harris (1971). *Men and dinosaurs: the search in field and laboratory*. Harmondsworth [Eng.]: Penguin. ISBN 0-14-021288-4.

[53] Lovelace, David M. (2007). "Morphology of a specimen of *Supersaurus* (Dinosauria, Sauropoda) from the Morrison Formation of Wyoming, and a re-evaluation of diplodocid phylogeny". *Arquivos do Museu Nacional* **65** (4): 527–544.

[54] dal Sasso C, Maganuco S, Buffetaut E, Mendez MA (2006). "New information on the skull of the enigmatic theropod *Spinosaurus*, with remarks on its sizes and affinities" (http://www.reocities.com/Athens/bridge/4602/spinoskull.pdf) (PDF). *Journal of Vertebrate Paleontology* **25** (4): 888–896. doi:10.1671/0272-4634(2005)025[0888:NIOTSO]2.0.CO;2. . Retrieved 2011-05-05.

[55] Therrien, F.; and Henderson, D.M. (2007). "My theropod is bigger than yours...or not: estimating body size from skull length in theropods". *Journal of Vertebrate Paleontology* **27** (1): 108–115. doi:10.1671/0272-4634(2007)27[108:MTIBTY]2.0.CO;2.

[56] Zhang F, Zhou Z, Xu X, Wang X, Sullivan C (2008). "A bizarre Jurassic maniraptoran from China with elongate ribbon-like feathers". *Nature* **455** (7216): 1105–1108. Bibcode 2008Natur.455.1105Z. doi:10.1038/nature07447. PMID 18948955.

[57] Xu X, Zhao Q, Norell M, Sullivan C, Hone D, Erickson G, Wang XL, Han FL, Guo Y. "A new feathered maniraptoran dinosaur fossil that fills a morphological gap in avian origin". *Chinese Science Bulletin* **54** (3): 430–435. doi:10.1007/s11434-009-0009-6.

[58] Rey LV, Holtz, Jr TR (2007). *Dinosaurs: the most complete, up-to-date encyclopedia for dinosaur lovers of all ages*. New York: Random House. ISBN 0-375-82419-7.

[59] Butler, R.J.; Zhao, Q. (2009). "The small-bodied ornithischian dinosaurs *Micropachycephalosaurus hongtuyanensis* and *Wannanosaurus yansiensis* from the Late Cretaceous of China". *Cretaceous Research* **30** (1): 63–77. doi:10.1016/j.cretres.2008.03.002.

[60] Yans J, Dejax J, Pons D, Dupuis C & Taquet P (2005). "Implications paléontologiques et géodynamiques de la datation palynologique des sédiments à faciès wealdien de Bernissart (bassin de Mons, Belgique)" (in French). *Comptes Rendus Palevol* **4** (1–2): 135–150. doi:10.1016/j.crpv.2004.12.003.

[61] Day, J.J.; Upchurch, P; Norman, DB; Gale, AS; Powell, HP (2002). "Sauropod trackways, evolution, and behavior". *Science* **296** (5573): 1659. doi:10.1126/science.1070167. PMID 12040187.

[62] Wright, Joanna L. (2005). "Steps in understanding sauropod biology". In Curry Rogers, Kristina A.; and Wilson, Jeffrey A.. *The Sauropods: Evolution and Paleobiology*. Berkeley: University of California Press. pp. 252–284. ISBN 0-520-24623-3.

[63] Varricchio, D.J.; Sereno, Paul C.; Xijin, Zhao; Lin, Tan; Wilson, Jeffery A.; Lyon, Gabrielle H. (2008). "Mud-trapped herd captures evidence of distinctive dinosaur sociality" (http://www.app.pan.pl/archive/published/app53/APP53-567.pdf) (PDF). *Acta Palaeontologica Polonica* **53** (4): 567–578. doi:10.4202/app.2008.0402. . Retrieved 2011-05-06.

[64] Lessem, Don; and Glut, Donald F. (1993). "*Allosaurus*". *The Dinosaur Society's Dinosaur Encyclopedia*. Random House. pp. 19–20. ISBN 0-679-41770-2.

[65] Maxwell, W. D.; Ostrom, John (1995). "Taphonomy and paleobiological implications of *Tenontosaurus–Deinonychus* associations". *Journal of Vertebrate Paleontology* **15** (4): 707–712. doi:10.1080/02724634.1995.10011256.(abstract (http://www.vertpaleo.org/publications/jvp/15-707-712.cfm))

[66] Roach, Brian T.; Brinkman, Daniel L. (2007). "A reevaluation of cooperative pack hunting and gregariousness in *Deinonychus antirrhopus* and other nonavian theropod dinosaurs". *Bulletin of the Peabody Museum of Natural History* **48** (1): 103–138. doi:10.3374/0079-032X(2007)48[103:AROCPH]2.0.CO;2.

[67] Horner, J.R ; Makela, Robert (1979). "Nest of juveniles provides evidence of family structure among dinosaurs". *Nature* **282** (5736): 296–298. Bibcode 1979Natur.282..296H. doi:10.1038/282296a0.

[68] Chiappe, Luis M.; Jackson, Frankie; Coria, Rodolfo A.; and Dingus, Lowell (2005). "Nesting titanosaurs from Auca Mahuevo and adjacent sites". In Curry Rogers, Kristina A.; and Wilson, Jeffrey A.. *The Sauropods: Evolution and Paleobiology*. Berkeley: University of California Press. pp. 285–302. ISBN 0-520-24623-3.

[69] "Discovering Dinosaur Behavior: 1960–present view" (http://search.eb.com/dinosaurs/dinosaurs/BRa.html+). Encyclopedia Brittanica. . Retrieved 2011-05-05.

[70] Meng Qingjin; Liu Jinyuan; Varricchio, David J.; Huang, Timothy; and Gao Chunling (2004). "Parental care in an ornithischian dinosaur". *Nature* **431** (7005): 145–146. Bibcode 2004Natur.431..145M. doi:10.1038/431145a. PMID 15356619.

[71] Reisz RR, Scott, D Sues, H-D, Evans, DC & Raath, MA (2005). "Embryos of an Early Jurassic prosauropod dinosaur and their evolutionary significance". *Science* **309** (5735): 761–764. Bibcode 2005Sci...309..761R. doi:10.1126/science.1114942. PMID 16051793.

[72] Clark NDL, Booth P, Booth CL, Ross DA (2004). "Dinosaur footprints from the Duntulm Formation (Bathonian, Jurassic) of the Isle of Skye" (http://testservice-eprints.gla.ac.uk/4496/1/4496.pdf) (PDF). *Scottish Journal of Geology* **40** (1): 13–21. . Retrieved 2011-05-05.

[73] Tanke, Darren H. (1998). "Head-biting behavior in theropod dinosaurs: paleopathological evidence" (http://www.mnhn.ul.pt/geologia/gaia/12.pdf) (PDF). *Gaia* (15): 167–184. ISSN 0871-5424. .

[74] "The Fighting Dinosaurs" (http://www.amnh.org/exhibitions/fightingdinos/ex-fd.html). American Museum of Natural History. . Retrieved 2007-12-05.

[75] Carpenter, K. (1998). "Evidence of predatory behavior by theropod dinosaurs" (http://vertpaleo.org/publications/jvp/15-576-591.cfm). *Gaia* **15**: 135–144. . Retrieved 2007-12-05.

[76] Rogers, Raymond R.; Krause, DW; Curry Rogers, K (2007). "Cannibalism in the Madagascan dinosaur *Majungatholus atopus*". *Nature* **422** (6931): 515–518. doi:10.1038/nature01532. PMID 12673249.

[77] Schmitz, L.; Motani, R. (2011). "Nocturnality in Dinosaurs Inferred from Scleral Ring and Orbit Morphology". *Science* **332** (6030): 705–708. Bibcode 2011Sci...332..705S. doi:10.1126/science.1200043. PMID 21493820.

[78] Varricchio DJ, Martin, AJ and Katsura, Y (2007). "First trace and body fossil evidence of a burrowing, denning dinosaur". *Proceedings of the Royal Society B: Biological Sciences* **274** (1616): 1361–1368. doi:10.1098/rspb.2006.0443. PMC 2176205. PMID 17374596.

[79] Chatterjee, S.; Templin, R. J. (2007). "Biplane wing planform and flight performance of the feathered dinosaur *Microraptor gui*" (http://www.pnas.org/cgi/reprint/0609975104v1.pdf) (PDF). *Proceedings of the National Academy of Sciences* **104** (5): 1576–1580. Bibcode 2007PNAS..104.1576C. doi:10.1073/pnas.0609975104. PMC 1780066. PMID 17242354. .

[80] Zhang, F.; Zhou, Z.; Xu, X.; and Wang, X. (2002). "A juvenile coelurosaurian theropod from China indicates arboreal habits". *Naturwissenschaften* **89** (9): 394–398. Bibcode 2002NW.....89..394Z. doi:10.1007/s00114-002-0353-8. PMID 12435090.

[81] Alexander RM (2006). "Dinosaur biomechanics". *Proceedings of the Royal Society of Biological Sciences* **273** (1596): 1849–1855. doi:10.1098/rspb.2006.3532. PMC 1634776. PMID 16822743.

[82] Goriely A & McMillen T (2002). "Shape of a cracking whip". *Physical Review Letters* **88** (24): 244301. Bibcode 2002PhRvL..88x4301G. doi:10.1103/PhysRevLett.88.244301. PMID 12059302.

[83] Henderson, D.M. (2003). "Effects of stomach stones on the buoyancy and equilibrium of a floating crocodilian: A computational analysis". *Canadian Journal of Zoology* **81** (8): 1346–1357. doi:10.1139/z03-122.

[84] Senter, P. (2008). "Voices of the past: a review of Paleozoic and Mesozoic animal sounds". *Historical Biology* **20** (4): 255–287. doi:10.1080/08912960903033327.

[85] Hopson, James A. (1975). "The evolution of cranial display structures in hadrosaurian dinosaurs". *Paleobiology* **1** (1): 21–43.

[86] Diegert, Carl F. (1998). "A digital acoustic model of the lambeosaurine hadrosaur *Parasaurolophus tubicen*". *Journal of Vertebrate Paleontology* **18** (3, Suppl.): 38A.

[87] Parsons, Keith M. (2001). *Drawing out Leviathan: Dinosaurs and the science wars*. Bloomington: Indiana University Press. pp. 22–48. ISBN 0-253-33937-5.

[88] Fisher, P. E., Russell, D. A., Stoskopf, M. K., Barrick, R. E., Hammer, M. & Kuzmitz, A. A. (2000). "Cardiovascular evidence for an intermediate or higher metabolic rate in an ornithischian dinosaur". *Science* **288** (5465): 503–505. Bibcode 2000Sci...288..503F. doi:10.1126/science.288.5465.503. PMID 10775107.

[89] Hillenius, W. J. & Ruben, J. A. (2004). "The evolution of endothermy in terrestrial vertebrates: Who? when? why?". *Physiological and Biochemical Zoology* **77** (6): 1019–1042. doi:10.1086/425185. PMID 15674773.

[90] Rowe T, McBride EF, & Sereno PC (2001). "Dinosaur with a Heart of Stone". *Science* **291** (5505): 783. doi:10.1126/science.291.5505.783a. PMID 11157158.

[91] Chinsamy A, Hillenius WJ (2004). "Physiology of nonavian dinosaurs". *The Dinosauria (2d edition*. University of California Press. pp. 643–659. ISBN 0-520-24209-2.

[92] Kaye, T. G.; Gaugler, G; Sawlowicz, Z; Stepanova, Anna (July 2008). "Dinosaurian Soft Tissues Interpreted as Bacterial Biofilms." (http://www.plosone.org/article/info:doi/10.1371/journal.pone.0002808). *PLoS ONE* **3** (7): e2808. Bibcode 2008PLoSO...3.2808K. doi:10.1371/journal.pone.0002808. PMC 2483347. PMID 18665236. . Retrieved 2008-10-27.

[93] "New Research Challenges Notion That Dinosaur Soft Tissues Still Survive" (http://newswise.com/articles/view/542898/). . Retrieved 2011-05-05.

[94] Wang, H., Yan, Z. and Jin, D. (1 May 1997). "Reanalysis of published DNA sequence amplified from Cretaceous dinosaur egg fossil" (http://mbe.oupjournals.org/cgi/reprint/14/5/589). *Molecular Biology and Evolution* **14** (5): 589–591. PMID 9159936. . Retrieved 2007-12-05.

[95] Chang BS, Jönsson K, Kazmi MA, Donoghue MJ, Sakmar TP (1 September 2002). "Recreating a Functional Ancestral Archosaur Visual Pigment" (http://mbe.oxfordjournals.org/cgi/content/full/19/9/1483). *Molecular Biology and Evolution* **19** (9): 1483–1489.

PMID 12200476. . Retrieved 2007-12-05.

[96] Schweitzer MH, Marshall M, Carron K, Bohle DS, Busse SC, Arnold EV, Barnard D, Horner JR, Starkey JR (1997). "Heme compounds in dinosaur trabecular bone". *Proc Natl Acad Sci U S A.* **94** (12): 6291–6. Bibcode 1997PNAS...94.6291S. doi:10.1073/pnas.94.12.6291. PMC 21042. PMID 9177210.

[97] Embery G, Milner AC, Waddington RJ, Hall RC, Langley MS, Milan AM (2003). "Identification of proteinaceous material in the bone of the dinosaur Iguanodon". *Connect Tissue Res* **44** (Suppl 1): 41–6. doi:10.1080/713713598. PMID 12952172.

[98] Huxley, Thomas H. (1868). "On the animals which are most nearly intermediate between birds and reptiles". *Annals of the Magazine of Natural History* **4** (2): 66–75.

[99] Heilmann, Gerhard (1926). *The Origin of Birds*. London: Witherby. pp. 208pp. ISBN 0-486-22784-7.

[100] Osborn, Henry Fairfield (1924). "Three new Theropoda, *Protoceratops* zone, central Mongolia" (http://digitallibrary.amnh.org/dspace/bitstream/2246/3223/1/N0144.pdf) (PDF). *American Museum Novitates* **144**: 1–12. .

[101] Ostrom, John H. (1973). "The ancestry of birds". *Nature* **242** (5393): 136. doi:10.1038/242136a0.

[102] Gauthier, Jacques. (1986). "Saurischian monophyly and the origin of birds". In Padian, Kevin. (ed.). *The Origin of Birds and the Evolution of Flight*. Memoirs of the California Academy of Sciences **8**. pp. 1–55.

[103] Mayr, G., Pohl, B. and Peters, D.S. (2005). "A Well-Preserved *Archaeopteryx* Specimen with Theropod Features". *Science* **310** (5753): 1483–1486. Bibcode 2005Sci...310.1483M. doi:10.1126/science.1120331. PMID 16322455.

[104] Martin, Larry D. (2006). "A basal archosaurian origin for birds". *Acta Zoologica Sinica* **50** (6): 977–990.

[105] Feduccia, A. (2002). "Birds are dinosaurs: simple answer to a complex problem". *The Auk* **119**: 1187–1201. doi:10.1642/0004-8038(2002)119[1187:BADSAT]2.0.CO;2.

[106] Wellnhofer, P (1988). "Ein neuer Exemplar von *Archaeopteryx*". *Archaeopteryx* **6**: 1–30.

[107] Xu X.; Norell, M.A.; Kuang X.; Wang X.; Zhao Q.; and Jia C. (2004). "Basal tyrannosauroids from China and evidence for protofeathers in tyrannosauroids". *Nature* **431** (7009): 680–684. Bibcode 2004Natur.431..680X. doi:10.1038/nature02855. PMID 15470426.

[108] Göhlich, U.B.; Chiappe, LM (2006). "A new carnivorous dinosaur from the Late Jurassic Solnhofen archipelago". *Nature* **440** (7082): 329–332. Bibcode 2006Natur.440..329G. doi:10.1038/nature04579. PMID 16541071.

[109] Lingham-Soliar, T. (2003). "The dinosaurian origin of feathers: perspectives from dolphin (Cetacea) collagen fibers". *Naturwissenschaften* **12** (12): 563–567. Bibcode 2003NW.....90..563L. doi:10.1007/s00114-003-0483-7. PMID 14676953.

[110] Feduccia, A.; Lingham-Soliar, T; Hinchliffe, JR (2005). "Do feathered dinosaurs exist? Testing the hypothesis on neontological and paleontological evidence". *Journal of Morphology* **266** (2): 125–166. doi:10.1002/jmor.10382. PMID 16217748.

[111] Lingham-Soliar, T.; Feduccia, A; Wang, X (2007). "A new Chinese specimen indicates that 'protofeathers' in the Early Cretaceous theropod dinosaur *Sinosauropteryx* are degraded collagen fibres". *Proceedings of the Biological Sciences* **274** (1620): 1823–9. doi:10.1098/rspb.2007.0352. PMC 2270928. PMID 17521978.

[112] Prum, Richard O. (April 2003). "Are Current Critiques Of The Theropod Origin Of Birds Science? Rebuttal To Feduccia 2002". *The Auk* **120** (2): 550–61. doi:10.1642/0004-8038(2003)120[0550:ACCOTT]2.0.CO;2. JSTOR 4090212.

[113] O'Connor, P.M. & Claessens, L.P.A.M. (2005). "Basic avian pulmonary design and flow-through ventilation in non-avian theropod dinosaurs". *Nature* **436** (7048): 253–256. Bibcode 2005Natur.436..253O. doi:10.1038/nature03716. PMID 16015329.

[114] Sereno, P.C.; Martinez, RN; Wilson, JA; Varricchio, DJ; Alcober, OA; Larsson, HC; Kemp, Tom (September 2008). "Evidence for Avian Intrathoracic Air Sacs in a New Predatory Dinosaur from Argentina" (http://www.plosone.org/article/info:doi/10.1371/journal.pone.0003303). *PLoS ONE* **3** (9): e3303. Bibcode 2008PLoSO...3.3303S. doi:10.1371/journal.pone.0003303. PMC 2553519. PMID 18825273. . Retrieved 2008-10-27.

[115] "Meat-Eating Dinosaur from Argentina Had Bird-Like Breathing System" (http://newswise.com/articles/view/544722/). . Retrieved 2011-05-05.

[116] Schweitzer, M.H.; Wittmeyer, JL; Horner, JR (2005). "Gender-specific reproductive tissue in ratites and *Tyrannosaurus rex*". *Science* **308** (5727): 1456–1460. Bibcode 2005Sci...308.1456S. doi:10.1126/science.1112158. PMID 15933198.

[117] Lee, Andrew H.; Werning, S (2008). "Sexual maturity in growing dinosaurs does not fit reptilian growth models" (http://www.pnas.org/cgi/content/abstract/105/2/582). *Proceedings of the National Academy of Sciences* **105** (2): 582–587. Bibcode 2008PNAS..105..582L. doi:10.1073/pnas.0708903105. PMC 2206579. PMID 18195356. .

[118] Xu, X. and Norell, M.A. (2004). "A new troodontid dinosaur from China with avian-like sleeping posture". *Nature* **431** (7010): 838–841. Bibcode 2004Natur.431..838X. doi:10.1038/nature02898. PMID 15483610.

[119] Norell M.A., Clark J.M., Chiappe L.M., Dashzeveg D. (1995). "A nesting dinosaur". *Nature* **378** (6559): 774–776. Bibcode 1995Natur.378..774N. doi:10.1038/378774a0.

[120] Varricchio, D. J.; Moore, J. R.; Erickson, G. M.; Norell, M. A.; Jackson, F. D.; Borkowski, J. J. (2008). "Avian Paternal Care Had Dinosaur Origin". *Science* **322**: 1826. Bibcode 2008Sci...322.1826V. doi:10.1126/science.1163245.

[121] Wings O (2007). "A review of gastrolith function with implications for fossil vertebrates and a revised classification" (http://www.app.pan.pl/archive/published/app52/app52-001.pdf) (PDF). *Palaeontologica Polonica* **52** (1): 1–16. . Retrieved 2011-05-05.

[122] Dingus, L. and Rowe, T. (1998). *The Mistaken Extinction – Dinosaur Evolution and the Origin of Birds*. New York: W. H. Freeman.

[123] Miller KG, Kominz MA, Browning JV, Wright JD, Mountain GS, Katz ME, Sugarman PJ, Cramer BS, Christie-Blick N, Pekar SF (2005). "The Phanerozoic record of global sea-level change". *Science* **310** (5752): 1293–8. Bibcode 2005Sci...310.1293M. doi:10.1126/science.1116412. PMID 16311326.

[124] McArthura JM, Janssenb NMM, Rebouletc S, Lengd MJ, Thirlwalle MF & van de Shootbruggef B (2007). "Palaeotemperatures, polar ice-volume, and isotope stratigraphy (Mg/Ca, δ18O, δ13C, 87Sr/86Sr): The Early Cretaceous (Berriasian, Valanginian, Hauterivian)". *Palaeogeography, Palaeoclimatology, Palaeoecology* **248** (3–4): 391–430. doi:10.1016/j.palaeo.2006.12.015.

[125] Alvarez, LW, Alvarez, W, Asaro, F, and Michel, HV (1980). "Extraterrestrial cause for the Cretaceous–Tertiary extinction". *Science* **208** (4448): 1095–1108. Bibcode 1980Sci...208.1095A. doi:10.1126/science.208.4448.1095. PMID 17783054.

[126] Hildebrand, Alan R.; Penfield, Glen T.; Kring, David A.; Pilkington, Mark; Zanoguera, Antonio Camargo; Jacobsen, Stein B.; Boynton, William V. (September 1991). "Chicxulub Crater; a possible Cretaceous/Tertiary boundary impact crater on the Yucatan Peninsula, Mexico". *Geology* **19** (9): 867–871. Bibcode 1991Geo....19..867H. doi:10.1130/0091-7613(1991)019<0867:CCAPCT>2.3.CO;2.

[127] Pope KO, Ocampo AC, Kinsland GL, Smith R (1996). "Surface expression of the Chicxulub crater". *Geology* **24** (6): 527–30. Bibcode 1996Geo....24..527P. doi:10.1130/0091-7613(1996)024<0527:SEOTCC>2.3.CO;2. PMID 11539331.

[128] Robertson, D.S.; et al. (30 September). "Survival in the first hours of the Cenozoic" (http://webh01.ua.ac.be/funmorph/raoul/macroevolutie/Robertson2004.pdf). *Geological Society of America Bulletin* **116** (5/6): 760–768. doi:10.1130/B25402.1. . Retrieved 15 June 2011.

[129] P, Claeys; Goderis, S (2007-09-05). "Solar System: Lethal billiards". *Nature* **449** (7158): 30–31. Bibcode 2007Natur.449...30C. doi:10.1038/449030a. PMID 17805281.

[130] Plotner, Tammy (2011). "Did Asteroid Baptistina Kill the Dinosaurs? Think other WISE..." (http://www.universetoday.com/89050/did-asteroid-baptistina-kill-the-dinosaurs-think-other-wise/#more-89050). *Universe Today*. . Retrieved 2011-9-20.

[131] edited by Christian Koeberl and Kenneth G. MacLeod. (2002). *Catastrophic Events and Mass Extinctions*. Geological Society of America. ISBN 0-8137-2356-6. OCLC 213836505.

[132] Hofman, C, Féraud, G & Courtillot, V (2000). "40Ar/39Ar dating of mineral separates and whole rocks from the Western Ghats lava pile: further constraints on duration and age of the Deccan traps". *Earth and Planetary Science Letters* **180**: 13–27. Bibcode 2000E&PSL.180...13H. doi:10.1016/S0012-821X(00)00159-X.

[133] Duncan, RA & Pyle, DG (1988). "Rapid eruption of the Deccan flood basalts at the Cretaceous/Tertiary boundary". *Nature* **333** (6176): 841–843. Bibcode 1988Natur.333..841D. doi:10.1038/333841a0.

[134] Alvarez, W (1997). *T. rex and the Crater of Doom*. Princeton University Press. pp. 130–146. ISBN 978-0-691-01630-6.

[135] Lloyd, G.T., Davis, K.E., Pisani, D. (22 July 2008). "Dinosaurs and the Cretaceous Terrestrial Revolution" (http://journals.royalsociety.org/content/7k63203q852h4006/). *Proceedings of the Royal Society: Biology* **275** (1650): 2483–90. doi:10.1098/rspb.2008.0715. PMC 2603200. PMID 18647715. . Retrieved 2008-07-28.

[136] Fassett, JE, Lucas, SG, Zielinski, RA, and Budahn, JR (2001). "Compelling new evidence for Paleocene dinosaurs in the Ojo Alamo Sandstone, San Juan Basin, New Mexico and Colorado, USA" (http://www.lpi.usra.edu/meetings/impact2000/pdf/3139.pdf) (PDF). *Catastrophic events and mass extinctions, Lunar and Planetary Contribution* **1053**: 45–46. . Retrieved 2007-05-18.

[137] Sloan, R. E., Rigby, K,. Van Valen, L. M., Gabriel, Diane (1986). "Gradual dinosaur extinction and simultaneous ungulate radiation in the Hell Creek Formation". *Science* **232** (4750): 629–633. Bibcode 1986Sci...232..629S. doi:10.1126/science.232.4750.629. PMID 17781415.

[138] Fastovsky, David E.; Sheehan, Peter M. (2005). "Reply to comment on "The Extinction of the dinosaurs in North America"" (http://www.geosociety.org/gsatoday/archive/15/7/pdf/i1052-5173-15-7-11b.pdf) (PDF). *GSA Today* **15**: 11. doi:10.1130/1052-5173(2005)015[11b:RTEOTD]2.0.CO;2. .

[139] Sullivan, RM (2003). "No Paleocene dinosaurs in the San Juan Basin, New Mexico" (http://gsa.confex.com/gsa/2003RM/finalprogram/abstract_47695.htm). *Geological Society of America Abstracts with Programs* **35** (5): 15. . Retrieved 2007-07-02.

[140] Fassett J.E., Heaman L.M., Simonetti A. (2011). "Direct U–Pb dating of Cretaceous and Paleocene dinosaur bones, San Juan Basin, New Mexico". *Geology* **39**: 159–162. doi:10.1130/G31466.1.

[141] Dong Zhiming (1992). *Dinosaurian Faunas of China*. China Ocean Press, Beijing. ISBN 3-540-52084-8. OCLC 26522845.

[142] "Dinosaur bones 'used as medicine'" (http://news.bbc.co.uk/2/hi/asia-pacific/6276948.stm). BBC News. 2007-07-06. . Retrieved 2007-07-06.

[143] Sarjeant WAS (1997). "The earliert discoveries". In Farlow JO, Brett-Surman MK. *The Complete Dinosaur*. Bloomington: Indiana University Press. pp. 3–11. ISBN 0-253-33349-0.

[144] Lhuyd, E. (1699). *Lithophylacii Britannici Ichnographia, sive lapidium aliorumque fossilium Britannicorum singulari figura insignium*. Gleditsch and Weidmann:London.

[145] Delair, J.B.; Sarjeant, W.A.S. (2002). "The earliest discoveries of dinosaurs: the records re-examined". *Proceedings of the Geologists' Association* **113**: 185–197.

[146] Gunther RT (1968). *Life and letters of Edward Lhwyd,: Second keeper of the Museum Ashmoleanum (Early science in Oxford Volume XIV)*. Dawsons of Pall Mall.

[147] Buckland W (1824). "Notice on the *Megalosaurus* or great Fossil Lizard of Stonesfield.". *Transactions of the Geological Society of London* **1**: 390–396.

[148] http://collections.tepapa.govt.nz/objectdetails.aspx?oid=212194&coltype=history&regno=gh004839

[149] Mantell, Gideon A. (1825). "Notice on the Iguanodon, a newly discovered fossil reptile, from the sandstone of Tilgate forest, in Sussex.". *Philosophical Transactions of the Royal Society* **115**: 179–186. doi:10.1098/rstl.1825.0010. JSTOR 107739.

[150] Sues, Hans-Dieter (1997). "European Dinosaur Hunters". In Farlow JO, Brett-Surman MK. *The Complete Dinosaur*. Bloomington: Indiana University Press. p. 14. ISBN 0-253-33349-0.

[151] Holmes T (1996). *Fossil Feud: The Bone Wars of Cope and Marsh, Pioneers in Dinosaur Science*. Silver Burdett Press. ISBN 978-0-382-39147-7. OCLC 34472600.

[152] Torrens, H.S. (1993). "The dinosaurs and dinomania over 150 years". *Modern Geology* **18** (2): 257–286.

[153] Breithaupt, Brent H. (1997). "First golden period in the USA". In Currie, Philip J. and Padian, Kevin (eds.). *The Encyclopedia of Dinosaurs*. San Diego: Academic Press. pp. 347–350. ISBN 978-0-12-226810-5.

[154] "*London. Michaelmas term lately over, and the Lord Chancellor sitting in Lincoln's Inn Hall. Implacable November weather. As much mud in the streets, as if the waters had but newly retired from the face of the earth, and it would not be wonderful to meet a Megalosaurus, forty feet long or so, waddling like an elephantine lizard up Holborne Hill.*" Dickens CJH (1852). *Bleak House*. London: Bradbury & Evans. p. 1.

[155] Glut, D.F., and Brett-Surman, M.K. (1997). Farlow, James O. and Brett-Surman, Michael K. (eds.). ed. *The Complete Dinosaur*. Indiana University Press. pp. 675–697. ISBN 978-0-253-21313-6.

Further reading

- Bakker, Robert T. (1986). *The Dinosaur Heresies: New Theories Unlocking the Mystery of the Dinosaurs and Their Extinction*. New York: Morrow. ISBN 0-688-04287-2.
- Holtz, Thomas R. Jr. (2007). *Dinosaurs: The Most Complete, Up-to-Date Encyclopedia for Dinosaur Lovers of All Ages*. New York: Random House. ISBN 978-0-375-82419-7.
- Paul, Gregory S. (2000). *The Scientific American Book of Dinosaurs*. New York: St. Martin's Press. ISBN 0-312-26226-4.
- Paul, Gregory S. (2002). *Dinosaurs of the Air: The Evolution and Loss of Flight in Dinosaurs and Birds*. Baltimore: The Johns Hopkins University Press. ISBN 0-8018-6763-0.

External links

General

- DinoDatabase.com | Hundreds of dinosaurs and dinosaur related topics. (http://www.dinodatabase.com/)

Images

- The Science and Art of Gregory S. Paul (http://www.gspauldino.com/) Influential paleontologist's anatomy art and paintings
- Skeletal Drawing (http://skeletaldrawing.com/) Professional restorations of numerous dinosaurs, and discussions of dinosaur anatomy.

Video

- BBC Nature: Watch dinosaurs brought to life and get experts' interpretations with videos from BBC programmes including Walking with Dinosaurs (http://www.bbc.co.uk/nature/life/dinosaur)

Popular

- Dinosaurs & other extinct creatures (http://www.nhm.ac.uk/nature-online/life/dinosaurs-other-extinct-creatures/index.html): From the Natural History Museum, a well illustrated dinosaur directory.
- Dinosaurnews (http://www.dinosaurnews.org/) (*www.dinosaurnews.org*) The dinosaur-related headlines from around the world. Recent news on dinosaurs, including finds and discoveries, and many links.
- Dinosauria (http://www.ucmp.berkeley.edu/diapsids/dinosaur.html) From UC Berkeley Museum of Paleontology Detailed information – scroll down for menu.
- LiveScience.com (http://www.livescience.com/dinosaurs/) All about dinosaurs, with current featured articles.
- Zoom Dinosaurs (http://www.enchantedlearning.com/subjects/dinosaurs/) (*www.enchantedlearning.com*) From Enchanted Learning. Kids' site, info pages and stats, theories, history.
- Dinosaur genus list (http://www.geol.umd.edu/~tholtz/dinoappendix/HoltzappendixWinter2010.pdf) contains data tables on nearly every published Mesozoic dinosaur genus as of January 2011.
- LiveScience.com (http://www.livescience.com/animals/090621-dinosaur-size.html) Giant Dinosaurs Get Downsized by *LiveScience*, June 21, 2009

Technical

- *Palaeontologia Electronica* (http://palaeo-electronica.org/) From Coquina Press. Online technical journal.
- Dinobase (http://dinobase.gly.bris.ac.uk/) A searchable dinosaur database, from the University of Bristol, with dinosaur lists, classification, pictures, and more.
- DinoData (http://www.dinodata.org/index.php) (*www.dinodata.org*) Technical site, essays, classification, anatomy.
- Dinosauria On-Line (http://www.dinosauria.com/dml/dml.htm) (*www.dinosauria.com*) Technical site, essays, pronunciation, dictionary.
- Thescelosaurus! (http://www.thescelosaurus.com/) By Justin Tweet. Includes a cladogram and small essays on each relevant genera and species.
- Dinosauromorpha Cladogram (http://www.palaeos.com/Vertebrates/Units/Unit310/000.html) From Palaeos (http://www.Palaeos.com/). A detailed amateur site about all things paleo.

Jurassic

Jurassic Period *199.6–145.5 million years ago*	
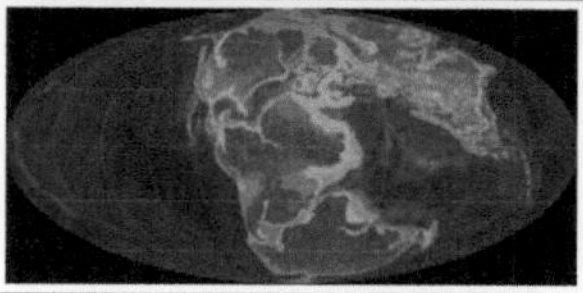	
Mean atmospheric O_2 content over period duration	ca. 26 Vol %[1] (130 % of modern level)
Mean atmospheric CO_2 content over period duration	ca. 1950 ppm[2] (7 times pre-industrial level)
Mean surface temperature over period duration	ca. 16.5 °C[3] (3 °C above modern level)

Key events in the Jurassic

view • discuss •

Cretaceous

Triassic

Jurassic

Hettangian

Sinemurian

Pliensbachian

Toarcian

Aalenian

Bajocian

Bathonian

Callovian

Oxfordian

Kimmeridgian

Tithonian

Mesozoic

An approximate timescale of key Jurassic events.
Axis scale: millions of years ago.

The **Jurassic** is a geologic period and system that extends from about 199.6± 0.6 Mya (million years ago) to 145.5± 4 Mya, that is, from the end of the Triassic to the beginning of the Cretaceous. The Jurassic constitutes the middle period of the Mesozoic Era, also known as the Age of Reptiles. The start of the period is marked by the major Triassic–Jurassic extinction event. However, the end of the Jurassic Period did not witness any major extinction event. The start and end of the period are defined by carefully selected locations; the uncertainty in dating arises from trying to date these horizons.

The chronostratigraphic term "Jurassic" is directly linked to the Jura Mountains. Alexander von Humboldt recognized the mainly limestone dominated mountain range of the Jura Mountains as a separate formation that was not at the time included in the established stratigraphic system defined by Abraham Gottlob Werner and named it "Jurakalk" in 1795.[4] [5] [6] The name "Jura" is derived from the celtic root "jor" which was Latinised into "juria", meaning forest (i.e. "Jura" is forest mountains).[4] [5] [7]

Divisions

The Jurassic Period is divided into Early Jurassic, Middle, and Late Jurassic epochs. The Jurassic System, in stratigraphy, is divided into Lower Jurassic, Middle, and Upper Jurassic series of rock formations, also known as *Lias*, *Dogger* and *Malm* in Europe.[8] The separation of the term **Jurassic** into three sections goes back to Leopold von Buch (* 1774, † 1853).[6] The faunal stages from youngest to oldest are:

Upper/Late Jurassic	
Tithonian	(150.8 ± 4.0 – 145.5 ± 4.0 Mya)
Kimmeridgian	(155.7 ± 4.0 – 150.8 ± 4.0 Mya)
Oxfordian	(161.2 ± 4.0 – 155.7 ± 4.0 Mya)
Middle Jurassic	
Callovian	(164.7 ± 4.0 – 161.2 ± 4.0 Mya)
Bathonian	(167.7 ± 3.5 – 164.7 ± 4.0 Mya)
Bajocian	(171.6 ± 3.0 – 167.7 ± 3.5 Mya)
Aalenian	(175.6 ± 2.0 – 171.6 ± 3.0 Mya)
Lower/Early Jurassic	
Toarcian	(183.0 ± 1.5 – 175.6 ± 2.0 Mya)
Pliensbachian	(189.6 ± 1.5 – 183.0 ± 1.5 Mya)
Sinemurian	(196.5 ± 1.0 – 189.6 ± 1.5 Mya)
Hettangian	(199.6 ± 0.6 – 196.5 ± 1.0 Mya)

Paleogeography and tectonics

Large dinosaurs roamed forests of similarly large conifers during the Jurassic Period.

During the early Jurassic period, the supercontinent Pangaea broke up into the northern supercontinent Laurasia and the southern supercontinent Gondwana; the Gulf of Mexico opened in the new rift between North America and what is now Mexico's Yucatan Peninsula. The Jurassic North Atlantic Ocean was relatively narrow, while the South Atlantic did not open until the following Cretaceous Period, when Gondwana itself rifted apart.[9] The Tethys Sea closed, and the Neotethys basin appeared. Climates were warm, with no evidence of glaciation. As in the Triassic, there was apparently no land near either pole, and no extensive ice caps existed.

The Jurassic geological record is good in western Europe, where extensive marine sequences indicate a time when much of the continent was submerged under shallow tropical seas; famous locales include the Jurassic Coast World Heritage Site and the renowned late Jurassic *lagerstätten* of Holzmaden and Solnhofen.[10] In contrast, the North American Jurassic record is the poorest of the Mesozoic, with few outcrops at the surface.[11] Though the epicontinental Sundance Sea left marine deposits in parts of the northern plains of the United States and Canada during the late Jurassic, most exposed sediments from this period are continental, such as the alluvial deposits of the Morrison Formation.

The Jurassic was a time of calcite sea geochemistry in which low-magnesium calcite was the primary inorganic marine precipitate of calcium carbonate. Carbonate hardgrounds were thus very common, along with calcitic ooids, calcitic cements, and invertebrate faunas with dominantly calcitic skeletons (Stanley and Hardie, 1998, 1999).

The first of several massive batholiths were emplaced in the northern Cordillera beginning in the mid-Jurassic, marking the Nevadan orogeny.[12] Important Jurassic exposures are also found in Russia, India, South America, Japan, Australasia and the United Kingdom.

Early Jurassic strata are distributed in a similar fashion to Late Triassic beds, with more common outcrops in the south and less common fossil beds which are predominated by tracks to the north.[13] As the Jurassic proceeded, larger and more iconic groups of dinosaurs like sauropods and ornithopods proliferated in Africa.[13] Middle Jurassic strata are neither well represented nor well studied in Africa.[13] Late Jurassic strata are also poorly represented apart from the spectacular Tendeguru fauna in Tanzania.[13] The Late Jurassic life of Tendeguru is very similar to that found in western North America's Morrison Formation.[13]

Jurassic limestones and marls (the Matmor Formation) in southern Israel.

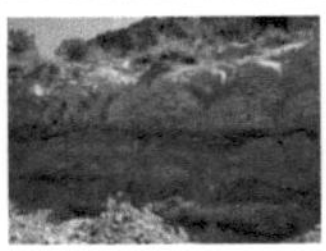
The late Jurassic Morrison Formation in Colorado is one of the most fertile sources of dinosaur fossils in North America.

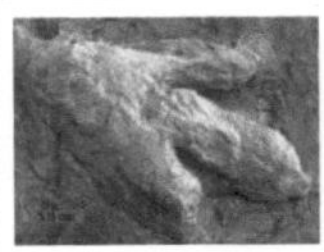
Gigandipus, a dinosaur footprint in the Lower Jurassic Moenave Formation at the St. George Dinosaur Discovery Site at Johnson Farm, southwestern Utah.

The Permian through Jurassic stratigraphy of the Colorado Plateau area of southeastern Utah.

Fauna

Aquatic and marine

During the Jurassic period, the primary vertebrates living in the seas were fish and marine reptiles. The latter include ichthyosaurs who were at the peak of their diversity, plesiosaurs, pliosaurs, and marine crocodiles of the families Teleosauridae and Metriorhynchidae.[14]

In the invertebrate world, several new groups appeared, including rudists (a reef-forming variety of bivalves) and belemnites. The Jurassic also had diverse encrusting and boring (sclerobiont) communities, and it saw a significant rise in the bioerosion of carbonate shells and hardgrounds. Especially common is the ichnogenus (trace fossil) *Gastrochaenolites*.[15]

During the Jurassic period about four or five of the twelve clades of planktonic organisms that exist in the fossil record either experienced a massive evolutionary radiation or appeared for the first time.[8]

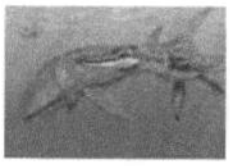

An over 10 metres long *Liopleurodon* (right) harassing an even larger *Leedsichthys* in a Jurassic sea.

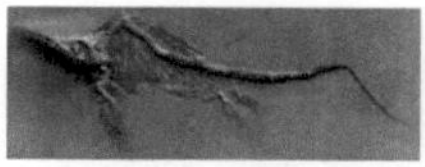

Ichthyosaurus from lower (early) Jurassic slates in southern Germany featured a dolphin-like body shape.

Plesiosaurs like *Muraenosaurus* roamed Jurassic oceans.

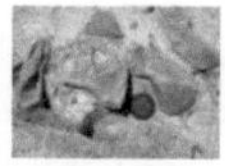

Gastropod and attached mytilid bivalves on a Jurassic limestone bedding plane in southern Israel.

Terrestrial

On land, large archosaurian reptiles remained dominant. The Jurassic was a golden age for the large herbivorous dinosaurs known as the sauropods—*Camarasaurus*, *Apatosaurus*, *Diplodocus*, *Brachiosaurus*, and many others—that roamed the land late in the period; their mainstays were either the prairies of ferns, palm-like cycads and bennettitales, or the higher coniferous growth, according to their adaptations. They were preyed upon by large theropods as for example *Ceratosaurus*, *Megalosaurus*, *Torvosaurus* and *Allosaurus*. All these belong to the 'lizard hipped' or saurischian branch of the dinosaurs.[16] During the Late Jurassic, the first birds, like Archaeopteryx, evolved from small coelurosaurian dinosaurs. Ornithischian dinosaurs were less predominant than saurischian dinosaurs, although some like stegosaurs and small ornithopods played important roles as small and medium-to-large (but not sauropod-sized) herbivores. In the air, pterosaurs were common; they ruled the skies, filling many ecological roles now taken by birds.[17] Within the undergrowth were various types of early mammals, as well as tritylodont mammal-like reptiles, lizard-like sphenodonts, and early lissamphibians.

The rest of the Lissamphibia evolved in this period, introducing the first salamanders and caecilians.[18]

Diplodocus reached lengths over 30 m was a common sauropod during the late Jurassic.

Allosaurus was one of the largest land predators during the Jurassic.

Stegosaurus is one of the most recognizable genera of dinosaurs and lived during the mid to late Jurassic.

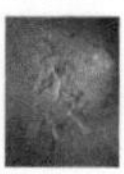

In the late Jurassic, the emergence of *Archaeopteryx* marks the start of the evolution of birds.

Flora

The arid, continental conditions characteristic of the Triassic steadily eased during the Jurassic period, especially at higher latitudes; the warm, humid climate allowed lush jungles to cover much of the landscape.[19] Gymnosperms were relatively diverse during the Jurassic period.[8] The Conifers in particular dominated the flora, as during the Triassic; they were the most diverse group and constituted the majority of large trees. Extant conifer families that flourished during the Jurassic included the Araucariaceae, Cephalotaxaceae, Pinaceae, Podocarpaceae, Taxaceae and Taxodiaceae.[20] The extinct Mesozoic conifer family Cheirolepidiaceae dominated low latitude vegetation, as did the shrubby Bennettitales.[21] Cycads were also common, as were ginkgos and Dicksoniaceous tree ferns in the forest.[8] Smaller ferns were probably the dominant undergrowth. Caytoniaceous seed ferns were another group of important plants during this time and are thought to have been shrub to small-tree sized.[22] Ginkgo plants were particularly common in the mid- to high northern latitudes.[8] In the Southern Hemisphere, podocarps were especially successful, while Ginkgos and Czekanowskiales were rare.[19] [21]

Conifers were the dominant land plants of the Jurassic period.

In the oceans modern coralline algae appeared for the first time.[8]

Notes

[1] Image:Sauerstoffgehalt-1000mj.svg
[2] Image:Phanerozoic Carbon Dioxide.png
[3] Image:All palaeotemps.png
[4] Hölder, H. 1964. Jura — Handbuch der stratigraphischen Geologie, IV. *Enke-Verlag*, 603 pp., 158 figs, 43 tabs; Stuttgart
[5] Arkell, W.J. 1956. Jurassic Geology of the World. *Oliver & Boyd*, 806 pp.; Edinburgh und London.
[6] Pieńkowski, G.; Schudack, M.E.; Bosák, P.; Enay, R.; Feldman-Olszewska, A.; Golonka, J.; Gutowski, J.; Herngreen, G.F.W.; Jordan, P.; Krobicki, M.; Lathuiliere, B.; Leinfelder, R.R.; Michalík, J.; Mönnig, E.; Noe-Nygaard, N.; Pálfy, J.; Pint, A.; Rasser, M.W.; Reisdorf, A.G.; Schmid, D.U.; Schweigert, G.; Surlyk, F.; Wetzel, A. & Theo E. Wong, T.E. 2008. Jurassic. In: McCann, T. (ed.): The Geology of Central Europe. Volume 2: Mesozoic and Cenozoic, *Geological Society*, pp.: 823-922; London.
[7] Rollier, L. 1903. Das Schweizerische Juragebirge. *Sonderabdruck aus dem Geographischen Lexikon der Schweiz, Verlag von Gebr. Attinger*, 39 pp; Neuenburg
[8] Kazlev, M. Alan (2002) Palaeos website (http://www.palaeos.com/Mesozoic/Jurassic/Jurassic.htm) Accessed July. 22, 2008
[9] Late Jurassic (http://www.scotese.com/late1.htm)
[10] Jurassic Period (http://www.urweltmuseum.de/Englisch/museum_eng/Geologie_eng/Tektonik_eng.htm)
[11] map (http://www.nationalatlas.gov/articles/geology/legend/ages/jurassic.html)
[12] Monroe and Wicander, 607.
[13] Jacobs, Louis, L. (1997). "African Dinosaurs". *Encyclopedia of Dinosaurs*. Edited by Phillip J. Currie and Kevin Padian. Academic Press. p. 2-4.
[14] Motani, R. (2000), Rulers of the Jurassic Seas, Scientific American vol.283, no. 6
[15] Taylor, P.D. and Wilson, M.A., 2003. Palaeoecology and evolution of marine hard substrate communities. Earth-Science Reviews 62: 1-103
[16] Haines, Tim (2000) Walking with Dinosaurs: A Natural History, New York: Dorling Kindersley Publishing, Inc.
[17] Feduccia, A. (1996). The Origin and Evolution of Birds. Yale University Press, New Haven. ISBN 0-300-06460-8
[18] Carroll, RL (1988), Vertebrate Paleontology and Evolution, WH Freeman & Co
[19] Haines, 2000.
[20] Behrensmeyer *et al.*, 1992, 349.
[21] Behrensmeyer *et al.*, 1992, 352
[22] Behrensmeyer *et al.*, 1992, 353

References

- Behrensmeyer, Anna K., Damuth, J.D., DiMichele, W.A., Potts, R., Sues, H.D. & Wing, S.L. (eds.) (1992), *Terrestrial Ecosystems through Time: the Evolutionary Paleoecology of Terrestrial Plants and Animals*, University of Chicago Press, Chicago and London, ISBN 0-226-04154-9 (cloth), ISBN 0-226-04155-7 (paper).
- Haines, Tim (2000) *Walking with Dinosaurs: A Natural History*, New York: Dorling Kindersley Publishing, Inc., p. 65. ISBN 0-563-38449-2.
- Kazlev, M. Alan (2002) Palaeos website (http://www.palaeos.com/Mesozoic/Jurassic/Jurassic.htm) Accessed Jan. 8, 2006.
- Mader, Sylvia (2004) Biology, eighth edition.
- Monroe, James S., and Reed Wicander. (1997) *The Changing Earth: Exploring Geology and Evolution*, 2nd ed. Belmont: West Publishing Company, 1997. ISBN 0-314-09577-2.
- Ogg, Jim; June, 2004, *Overview of Global Boundary Stratotype Sections and Points (GSSP's)*, International Commission on Stratigraphy, pp. 17
- Stanley, S.M. and Hardie, L.A. (1998). "Secular oscillations in the carbonate mineralogy of reef-building and sediment-producing organisms driven by tectonically forced shifts in seawater chemistry". Palaeogeography, Palaeoclimatology, Palaeoecology 144: 3–19.
- Stanley, S.M. and Hardie, L.A. (1999). "Hypercalcification; paleontology links plate tectonics and geochemistry to sedimentology". GSA Today 9: 1–7.
- Taylor, P.D. and Wilson, M.A., 2003. Palaeoecology and evolution of marine hard substrate communities. Earth-Science Reviews 62: 1–103. (http://www3.wooster.edu/geology/Taylor&Wilson2003.pdf).

External links

- Examples of Jurassic Fossils (http://www.geo-lieven.com/erdzeitalter/jura/jura.htm)
- Palaeos.com (http://www.palaeos.com/Mesozoic/Jurassic/Jurassic.htm)
- Jurassic fossils in Harbury, Warwickshire (http://harbury.villagebuzz.co.uk/viewtopic.php?f=16&t=297)
- Jurassic Microfossils: 65+ images of Foraminifera (http://www.foraminifera.eu/querydb.php?period=Jurassic&aktion=suche)

Jurassic Period		
Lower/Early Jurassic	**Middle Jurassic**	**Upper/Late Jurassic**
Hettangian | Sinemurian Pliensbachian | Toarcian	Aalenian | Bajocian Bathonian | Callovian	Oxfordian | Kimmeridgian Tithonian

Preceded by Proterozoic Eon	542 Ma - Phanerozoic Eon - Present											
	542 Ma - Paleozoic Era - 251 Ma						251 Ma - Mesozoic Era - 65 Ma			65 Ma - Cenozoic Era - Present		
	Cambrian	Ordovician	Silurian	Devonian	Carboniferous	Permian	Triassic	Jurassic	Cretaceous	Paleogene	Neogene	*Quaternary*

Ornithopod

Ornithopods Temporal range: Early Jurassic - Late Cretaceous,	

Various ornithopod dinosaurs. Far left: *Camptosaurus*, left: *Iguanodon*, centre background: *Shantungosaurus*, centre foreground: *Dryosaurus*, right: *Corythosaurus*, far right (small): *Heterodontosaurus* (probably not an ornithopod), far right: *Tenontosaurus*.	
Scientific classification	
Kingdom:	Animalia
Phylum:	Chordata
Class:	Reptilia
Superorder:	Dinosauria
Order:	†Ornithischia
Node:	†Cerapoda
Suborder:	**†Ornithopoda** Marsh, 1881
Families	
• Camptosauridae • Dryosauridae • Hadrosauridae • Hypsilophodontidae* • Iguanodontidae • Rhabdodontidae	

Ornithopods (pronounced /ɔrˈnɪθəpɒd/) or members of the clade **Ornithopoda** (pronounced /ɔrnɨˈθɒpədə/) are a group of bird-hipped dinosaurs that started out as small, bipedal running grazers, and grew in size and numbers until they became one of the most successful groups of herbivores in the Cretaceous world, and dominated the North American landscape. Their major evolutionary advantage was the progressive development of a chewing apparatus that became the most sophisticated ever developed by a reptile, rivaling that of modern mammals like the domestic cow. They reached their apex in the duck-bills, before they were wiped out by the Cretaceous-Tertiary extinction event along with all other non-avian dinosaurs. Members are known from all seven continents, although the Antarctic remains are unnamed, and they are generally rare in the Southern Hemisphere.

Description

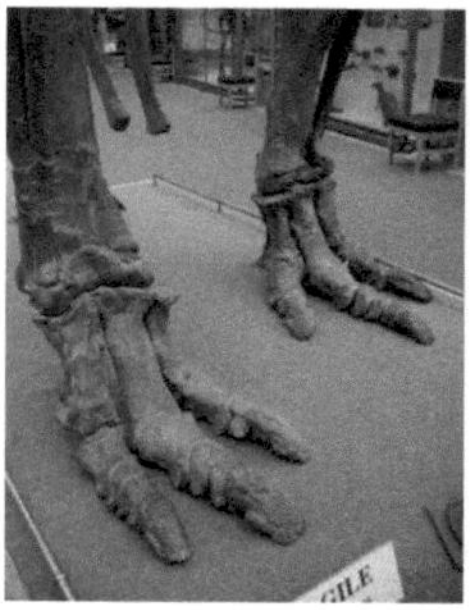
Three-toed feet of *Iguanodon*

Ornithopoda means "bird feet", from the Greek *ornithos* ("bird") and *pous* ("feet"); this refers to their characteristic three-toed feet, although many early forms retained four toes. They were also characterized by having no armor, the development of a horny beak, an elongated pubis that eventually extended past the ilium, and a missing hole in the lower jaw. A variety of ornithopods and related cerapods had thin cartilaginous plates along the outside of the ribs; in some cases, these plates mineralized and were fossilized. The function of these intercostal plates is unknown. They have been found with *Hypsilophodon*, *Othnielosaurus*, *Parksosaurus*, *Talenkauen*, *Thescelosaurus*,[1] and *Macrogryphosaurus* to date.[2]

The early ornithopods were only about 1 meter (3 feet) long, but probably very fast. They had a stiff tail, like the theropods, to help them balance as they ran on their hind legs. Later ornithopods became more adapted to grazing on all fours; their spines curved, and came to resemble the spines of modern ground-feeders like the bison. As they became more adapted to eating while bent over, they became semi-quadrupedal; still running on two legs, and comfortable reaching up into trees; but spending most of their time walking or grazing while on all fours. The taxonomy of dinosaurs previously ascribed to the hypsilophodontidae is problematic. The group previously consisted of all non-Iguanodontid bipedal ornithischians, but a phylogenetic reappraisal has shown such species to be paraphyletic. As such, the hypsilophodont family is represented only by *Hypsilophodon*.[3]

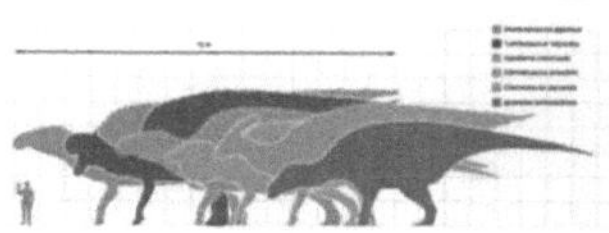
Sizes of the largest ornithopods

Later ornithopods became larger, but never rivaled the incredible size of the long-necked, long-tailed sauropods that they partially supplanted. The very largest, like *Shantungosaurus* were as heavy as medium sized sauropods at up to 23 metric tons (25 short tons) but never grew far beyond 15 meters (50 feet).

Classification

Historically, most indeterminate ornithischian bipeds were lumped in as ornithopods. Most have since been reclassified as basal members of typically quadrupedal taxa like Marginocephalia; and some, like the "bone-headed" pachycephalosaurids, have been given their own taxa.

Artist's impression of several species in profile

Taxonomy

Ornithopoda is usually given the rank of Suborder, within the order Ornithischia. While ranked taxonomy has largely fallen out of favor among dinosaur paleontologists, some researchers have continued to employ such a classification, though sources have differed on what its rank should be. Benton (2004) placed it as an infraorder within the suborder Cerapoda (originally named as an unranked clade), while others, such as Ibiricu *et al.* 2010, have retained it at its traditional rank of suborder.[4]

Suborder Ornithopoda

- *Changchunsaurus*
- Family Hypsilophodontidae*
- **Unranked clade Iguanodontia**
  - *Anabisetia*
  - ?*Bihariosaurus*
  - **Elasmaria**
    - *Macrogryphosaurus*
    - *Talenkauen*
  - *Tenontosaurus*
  - Family Camptosauridae
  - Family Dryosauridae
  - Family Rhabdodontidae
  - **Hadrosauriformes**
    - Family Iguanodontidae
    - **Superfamily Hadrosauroidea**
      - Family Hadrosauridae

Phylogeny

The cladogram below follows a 2009 analysis by Zheng and colleagues.[5]

Othnielia

Hypsilophodon

Jeholosaurus

Yandusaurus

Orodromeus

Zephyrosaurus

Marginocephalia

Pachycephalosauria

Ceratopsia

Anabisetia

Bugenasaura

Thescelosaurus

Cerapoda

Talenkauen

Gasparinisaura

Parksosaurus

Ornithopoda

Rhabdodontidae

Tenontosaurus

T. tilletti

T. dossi

Iguanodontia

Dryosauridae

Ankylopollexia

Cladogram after Butler *et al*, 2011.[6]

Marginocephalia

Pachycephalosauria

Ceratopsia

Orodromeus

Hypsilophodon

Zephyrosaurus

Yandusaurus

Changchunsaurus

Jeholosaurus

Gasparinisaura

Parksosaurus

Bugenasaura

Cerapoda

Thescelosaurus

Ornithopoda

Talenkauen

Anabisetia

Rhabdodontidae

T. tilletti

Iguanodontia *Tenontosaurus*

T. dossi

Dryosauridae

Ankylopollexia

References

[1] Butler, Richard J.; and Galton, Peter M. (2008). "The 'dermal armour' of the ornithopod dinosaur *Hypsilophodon* from the Wealden (Early Cretaceous: Barremian) of the Isle of Wight: a reappraisal". *Cretaceous Research* **29** (4): 636–642. doi:10.1016/j.cretres.2008.02.002.

[2] Calvo, J.O.; Porfiri, J.D.; and Novas, F.E. (2007). "Discovery of a new ornithopod dinosaur from the Portezuelo Formation (Upper Cretaceous), Neuquén, Patagonia, Argentina.". *Arquivos do Museu Nacional* **65** (4): 471–483.

[3] Weishampel, D.B., and Heinrich, R.E. (1992). "Systematics of Hypsilophodontidae and basal Iguanodontia (Dinosauria: Ornithopoda)". *Historical Biology* **6** (3): 159–184. doi:10.1080/10292389209380426.

[4] Ibiricu, L.M., Martínez, R.D., Lamanna, M.C., Casal, G.A., Luna, M., Harris, J.D. and Lacovara, K.J. (2010). "A Medium-Sized Ornithopod (Dinosauria: Ornithischia) from the Upper Cretaceous Bajo Barreal Formation of Lago Colhué Huapi, Southern Chubut Province, Argentina." *Annals of Carnegie Museum*, **79**(1): 39-50. doi:10.2992/007.079.0103

[5] Zheng, Xiao-Ting; You, Hai-Lu; Xu, Xing; Dong, Zhi-Ming (19 March 2009). "An Early Cretaceous heterodontosaurid dinosaur with filamentous integumentary structures". *Nature* **458** (7236): 333–336. doi:10.1038/nature07856. PMID 19295609.

[6] Richard J. Butler, Jin Liyong, Chen Jun, Pascal Godefroit (2011). "The postcranial osteology and phylogenetic position of the small ornithischian dinosaur *Changchunsaurus parvus* from the Quantou Formation (Cretaceous: Aptian–Cenomanian) of Jilin Province, north-eastern China". *Palaeontology* **54** (3): 667–683. doi:10.1111/j.1475-4983.2011.01046.x.

External links

- Ornithopoda (http://www.thescelosaurus.com/ornithopoda.htm), by Justin Tweet, from *Thescelosaurus*!
- GEOL 104 dinosaurs: a Natural history, ornithopoda: beaks, bills & crests (http://www.geol.umd.edu/~tholtz/G104/10419orni.htm), by Thomas R. Holtz, Jr.

Miguel Telles Antunes

Miguel Telles Antunes (born January 11, 1937 Lisbon) is a Portuguese palaeontologist. He is specializes, among other things, in dinosaur palaeontology. Antunes is the director of the Museu Maynense da Academia das Ciências de Lisboa.[1] He is currently associated with many institutions, including the Faculdade de Ciência e Tecnologia of the Universidade Nova de Lisboa, the Academia das Ciências de Lisboa and the Museu da Lourinhã.

The theropod dinosaur Lourinhanosaurus antunesi was dedicated to him.

Publications (selected)

ANTUNES, M.T. (1976) Dinossáurios eocretácicos de Lagosteiros, Ciências da Terra 1:1-35.

ANTUNES, M.T. (1984). Novas pistas de Dinossáurios no cretácico inferior- Discussão. Comunicações dos Serviços Geológicos de Portugal. 70(1): 123-4.

ANTUNES, M.T. (1986). Sobre a história da Paleontologia em Portugal. Memórias da Academia de Ciências de Lisboa. II: 773-814.

ANTUNES, M. T. (1992) - Sobre a História da Paleontologia em Portugal (ca. 1919-1980). História e Desenvolvimento da Ciência em Portugal no séc. XX . Publicações do II Centenário da Academia das Ciências de Lisboa, p. 1003–1026, 18 fig.

Antunes, M. Telles, (1989). Sobre a História do Ensino da Geologia em Portugal. Comunicações dos Serviços Geológicos de Portugal, 75: 127-160.

ANTUNES, M.T. (1998). A new Upper Jurassic Paulchoffatiid Multituberculate (Mammalia) from Pai Mogo, Portugal / and a few comments on Walter Georg Kühne. Memórias da Academia de Ciências de Lisboa. 37:125-153.

ANTUNES, M.T. (1999). Dinossauros e Portugal: Dois casos menos conhecidos. Ciências da Terra 13.

ANTUNES, M.T. (1999). Veiga Ferreira e a Paleontologia em Portugal. Ciências da Terra (UNL), No. 13:157-167.

ANTUNES, M. Telles (2000) Paleontologia e Portugal. Colóquio/ Ciência/ Revista de Cultura Científica, 25: 54 - 75, 50 fig. Fundação Calouste Gulbenkian, Lisboa.

ANTUNES, M.T. (2001) The earliest illustration of Dinosaur footprints. Proceedings of the INHIGEO Meeting, Portugal, 2001. XXX

ANTUNES, M.T. & SIGOGNEAU-RUSSELL, D. (1991). Nouvelles données sur les Dinosaures du Crétacé supérieur du Portugal. C.R. Acad. Sci. Paris. 313(II): 113-119.

ANTUNES, M.T. & SIGOGNEAU-RUSSELL, D. (1992). La faune de petits dinosaures du Crétacé terminal portugais. Comunicações dos Serviços Geológicos de Portugal, 78(1): 49-62.

ANTUNES, M.T. & SIGOGNEAU-RUSSELL, D. (1995). O Cretácico terminal português e o seu contributo para o esclarecimento da extinção dos dinossauros. Memórias da Academia das Ciências de Lisboa. 35: 131-144.

ANTUNES, M.T. & SIGOGNEAU-RUSSELL, D. (1996). Le Crétacé terminal portugais et son apport au problème de l'extinction des dinosaures. Bulletin du Muséum National d'Histoire Naturelle, 4 ser., 18(4): 595-606.

MATEUS, O. & ANTUNES, M.A. 2003. A New Dinosaur Tracksite in the Lower Cretaceous of Portugal. Ciências da Terra (UNL), 15: 253-262.

ANTUNES, M.T. & MATEUS, O. 2003. Dinosaurs of Portugal. C. R. Palevol, 2: 77-95.

MATEUS, O. & ANTUNES, M.T. 2001. Draconyx loureiroi, a new Camptosauridae (Dinosauria: Ornithopoda) from the Late Jurassic of Lourinhã, Portugal. Annales de Paleontologie, 87(1): 61-73.

Ricqles, A. DE, Mateus, O., Antunes, M. Telles & Taquet, P. 2001. Histomorphogenesis of embryos of Upper Jurassic Theropods from Lourinhã (Portugal). Comptes rendus de l'Académie des sciences - Série IIa - Sciences de la Terre et des planètes. 332(10): 647-656.

MATEUS, O., ANTUNES, M.T. & TAQUET, P. 2001. Dinosaur ontogeny: The case of Lourinhanosaurus (Late Jurassic, Portugal). Journal of Vertebrate Paleontology, 21 (Suppl. 3): 78A.

MATEUS, I., MATEUS, H., ANTUNES, M.T, MATEUS, O., TAQUET, P. RIBEIRO, V. & MANUPPELLA, G. 1997. Upper jurassic theropod dinosaur embryos from Lourinhã (Portugal). Memórias da Academia de Ciências de Lisboa. 37: 101-110.

MATEUS, I., MATEUS, H., TELLES ANTUNES, M., MATEUS, O., TAQUET, P., RIBEIRO, V. & MANUPPELLA, G. 1997. Couvée, œufs et embryons d'un Dinosaure Théropode du Jurassique de Lourinhã (Portugal). C.R Acad. Sci. Paris, Sciences de la terre et des planètes, 325: 71-78.

References

[1] "Mensagem do Director do Museu da Academia das Ciências de Lisboa" (http://www.acad-ciencias.pt/index.php?option=com_content&view=article&id=97&Itemid=80). . Retrieved 2009-05-10.

Octávio Mateus

Octávio Mateus (born in 1975) is a Portuguese dinosaur paleontologist and biologist (graduated in Universidade de Évora). He received his PhD at Universidade Nova de Lisboa in 2005. He collaborates with Museu da Lourinhã, known for their dinosaur collection.

Specialist in dinosaurs, he has studied Late Jurassic dinosaurs of Portugal, publishing several scientific articles.

He has named new dinosaur species such as *Lourinhanosaurus antunesi* [1] (1998), *Dinheirosaurus lourinhanensis*,[2] *Tangvayosaurus hoffeti* (1999), *Draconyx loureiroi* (2001) , *Lusotitan atalaiensis* (2003), *Europasaurus holgeri* (2006), and *Allosaurus europaeus* (2006).

Since 1991 Octávio Mateus has organized dinosaur excavations in Portugal, as well as excavating in Laos (Asian Southeast) with the French team of the Paris Museum of Natural History, led by Prof. Philippe Taquet. He has recently worked in Angola, where he discovered the first Angolan dinosaur in the scope of a project in the area of vertebrate paleontology of Angola.[3]

He collaborates with diverse international scientific institutions as the scientific council member of the German foundation Verein zur Förderung der niedersächsischen Paläontologie.

Works

- Antunes M.T., Mateus O. (2003). "Dinosaurs of Portugal". *C. R. Palevol* **2**: 77–95. doi:10.1016/S1631-0683(03)00003-4.
- Mateus O (1998). "*Lourinhanosaurus antunesi*, a new Upper Jurassic allosauroid (Dinosauria: Theropoda) from Lourinhã (Portugal)" [4] (PDF). *Memórias da Academia de Ciências de Lisboa* **37**: 111–124.

References

[1] Benson, R.B.J., Carrano, M.T and Brusatte, S.L. (2010). "A new clade of archaic large-bodied predatory dinosaurs (Theropoda: Allosauroidea) that survived to the latest Mesozoic". *Naturwissenschaften* **97** (1): 71–78. Bibcode 2010NW.....97...71B. doi:10.1007/s00114-009-0614-x. PMID 19826771. Supporting Information (http://www.springerlink.com/content/l496325vp2x32617/MediaObjects/114_2009_614_MOESM1_ESM.pdf)

[2] Bonaparte, J.F., & Mateus, O. (1999). "A new diplodocid, *Dinheirosaurus lourinhanensis* gen. et sp. nov., from the Late Jurassic beds of Portugal". *Revista del Museo Argentino de Ciencias Naturales* (2): 13–29. PDF Document (http://www.museulourinha.org/pt/Omateus/Papers/BonaparteMateus1999.pdf)

[3] Huffington Post (http://www.huffingtonpost.com/2011/03/16/angolatitan-adamastor-dinosaur-discovered_n_836647.html) - Angolatitan Adamastor: New Dinosaur, 'Angolan Giant,' Discovered

[4] http://museulourinha.org/pt/omateus/papers/mateus%201998%20lourinhanosaurus.pdf

Lourinhã Municipality

Lourinhã	
— Municipality —	
 Flag	 Coat of arms
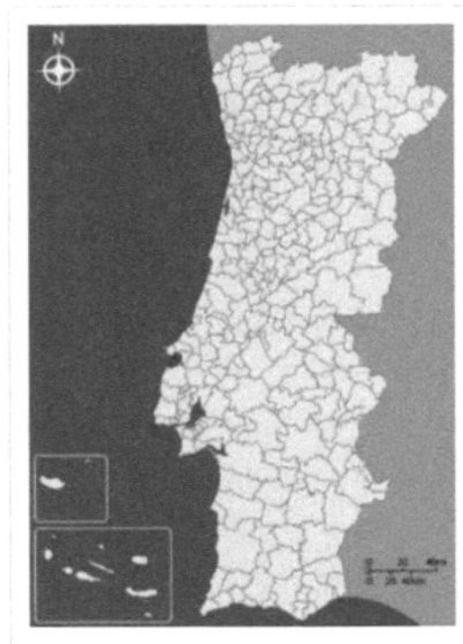 Location in Portugal	
Coordinates: 39°14′N 9°18′W	
Country	Portugal
Region	Centro, Portugal
Subregion	Oeste
District/A.R.	Lisbon
Government	
- Mayor	José Custódio (PS)
Area	
- Total	147.2 km² (56.8 sq mi)
Population	
- Total	24,601
- Density	167/km² (432.5 sq mi)
Parishes (no.)	11
Municipal holiday	Saint John June 24
Website	http://www.cm-lourinha.pt

Lourinhã (Portuguese pronunciation: [loɾiˈɲɐ̃]) is a municipality in Portugal with a total area of 147.2 km² and a total population of 24,601 inhabitants. The seat of the municipality is the village of Lourinhã, with a population of 8,800 inhabitants.

The municipality is composed of 11 parishes, and is located in the District of Lisbon.

The present Mayor is José Manuel Dias Custódio, elected by the Socialist Party.

History

The name Lourinhã possibly originated in the period of Roman domination, when a villa was located in the area. The origin of the medieval village is linked to *Jordan*, a French knight who took part in the successful Siege of Lisbon in 1147. King Afonso Henriques granted Jordan the region of Lourinhã as fief and allowed him to grant a foral (letter of feudal rights) to its settlers in 1160. The name Lourinhã may be related to the origin of its feudal lord, since Jordan was from the Loire region in France.

Santo António Convent Church in Lourinhã.

The rights of Lourinhã were confirmed by letters of King Sancho I in 1218 and again by Afonso III in 1258. The parish of Lourinhã became one of the richest of the Lisbon Diocese, as reflected by its main church, a fine example of 14th-century Portuguese Gothic architecture. The Gothic works of the main church were sponsored by Lourenço Vicente, a Lourinhã-born Archbishop of Braga who received the village as a donation of King John I in 1384.

In the 16th century, the Franciscan monastery of Santo António (1598) and a Misericórdia Church and Charity (1586) were founded in Lourinhã. The Misericórdia (Mercy), a religious charity, now houses a museum with outstanding Renaissance paintings. The most important paintings are by the hand of a mysterious early 16th-century painter, dubbed the *Master of Lourinhã* (Mestre da Lourinhã).

Starting at the end of the 19th century, the infrastructure of the municipality was modernised with roads, canalised water and electric light, as well as improvements in the educational system. The economy depended mostly on agriculture and fishing. Tourism is an increasingly important source of revenue, due to the municipality's extensive, picturesque beaches and, more recently, by the paleontological remains, which include fossilised bones, footprints, eggs and even embryos from Jurassic dinosaurs. Many of which can be seen nowadays at the local museum, Museu da Lourinhã.

Lourinhã, is one of the few brandy-making areas, besides Cognac, Armagnac and Jerez, that have received appellation status.

Civil Parishes

- Atalaia
- Lourinhã (parish)
- Marteleira
- Miragaia
- Moita dos Ferreiros
- Moledo
- Reguengo Grande
- Ribamar
- Santa Bárbara
- São Bartolomeu dos Galegos
- Vimeiro

Population

Population of Lourinhã municipality (1801 – 2004)								
1801	1849	1900	1930	1960	1981	1991	2001	2004
3366	6481	12154	17049	22927	21245	21596	23265	24601

Beaches

Lourinhã has 12 km of coastline with several popular beaches.

- Praia de Areal
- Praia da Areia Branca
- Praia do Caniçal
- Praia de Paimogo
- Praia de Vale Frades
- Praia de Porto das Barcas
- Praia da Peralta

Areia Branca beach.

Paleontology

The area of Lourinhã is known by the Late Jurassic findings of dinosaurs and other fossils, and give the name for Lourinhã Formation. The Museu da Lourinhã holds the main dinosaur collection.

References and external links

- Municipality official website [1]

References

[1] http://www.cm-lourinha.pt

Portugal

Portuguese Republic
***República Portuguesa* (Portuguese)**

Anthem: "*A Portuguesa*"
"The Portuguese Anthem"

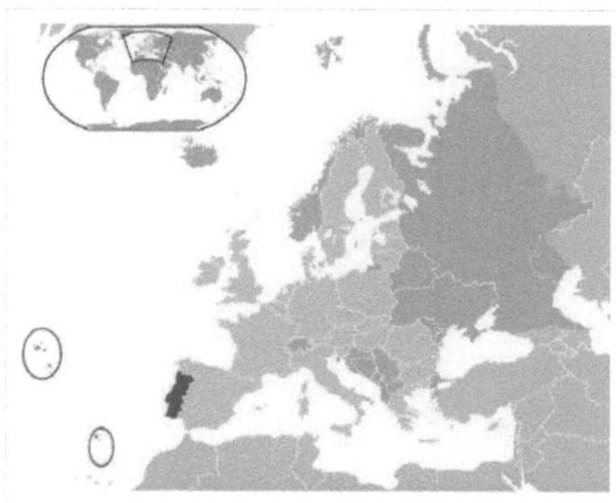

Location of Portugal(dark green)

– in Europe(green & dark grey)
– in the European Union(green) — [Legend]

Capital (and largest city)		Lisbon 38°46′N 9°9′W
Official language(s)		Portuguese
Recognised regional languages		Mirandese[1]
Ethnic groups (2007)		96.87% Portuguese and 3.13% legal immigrants (Cape Verdeans, Brazilians, Ukrainians, Angolans, etc.)[1]
Demonym		Portuguese
Government		Parliamentary republic
-	President	Aníbal Cavaco Silva (PSD)
-	Prime Minister	Pedro Passos Coelho (PSD)
-	Assembly President	Assunção Esteves (PSD)
Formation		Conventional date for Independence is 1139
-	Founding	868
-	Re-founding	1095
-	*De facto* sovereignty	24 June 1128

-	Kingdom	25 July 1139
-	Recognized	5 October 1143
-	Papal Recognition	23 May 1179
-	Restoration of independence	1 December 1640
-	Restoration of independence recognized	13 February 1668
-	Republic	5 October 1910
-	Democracy	25 April 1974
	Area	
-	Total	92,090 km^2 (110th) 35,645 sq mi
-	Water (%)	0.5
	Population	
-	2011 estimate	10,647,763[2] (77th)
-	2011 census	10,555,853[3]
-	Density	115/km^2 (96th) 298/sq mi
GDP (PPP)		2010 estimate
-	Total	$247.037 billion[4]
-	Per capita	$23,222[4]
GDP (nominal)		2010 estimate
-	Total	$229.336 billion[4]
-	Per capita	$21,558[4]
Gini (2009)		33.7[5]
HDI (2010)		▲ 0.795[6] (very high) (40th)
Currency		Euro (€)^2 (EUR)
Time zone		WET (UTC+0)
-	Summer (DST)	WEST (UTC+1)
Date formats		dd-mm-yyyy, yyyy-mm-dd, yyyy/mm/dd
Drives on the		right
ISO 3166 code		PT
Internet TLD		.pt
Calling code		351
1	Mirandese, spoken in some villages of the municipality of Miranda do Douro, was officially recognized in 1999 (*Lei n.° 7/99 de 29 de Janeiro*), since then awarding an official right-of-use Mirandese to the linguistic minority it is concerned.[7] The Portuguese Sign Language is also recognized.	
2	Before 1999: Portuguese escudo.	

Portugal i/ pɔrtʃʉgəl/ (Portuguese: *Portugal*, IPA: [purtuˈɣaɫ]), officially the **Portuguese Republic** (Portuguese: *República Portuguesa*) is a country situated in southwestern Europe on the Iberian Peninsula. Portugal is the westernmost country of Europe, and is bordered by the Atlantic Ocean to the West and South and by Spain to the North and East. The Atlantic archipelagos of the Azores and Madeira are part of Portugal. The country is named after its second largest city, Porto, whose Latin name was *Portus Cale*.[8]

The land within the borders of today's Portuguese Republic has been continuously settled since prehistoric times: occupied by Celts like the Gallaeci and the Lusitanians, integrated into the Roman Republic and later settled by Germanic peoples such as the Suebi and the Visigoths, in the 8th century the lands were conquered by Moors. During the Christian *Reconquista*, Portugal established itself as an independent kingdom from León, claiming to be the oldest European nation-state.[9]

In the 15th and 16th centuries, as the result of pioneering the Age of Discovery, Portugal expanded western influence and established a global empire that included possessions in Africa, Asia, Oceania, and South America, becoming one of the world's major economic, political and military powers. The Portuguese Empire was the first transoceanic empire in history,[10] and also the longest lived of the European colonial empires. However, the country's international status was greatly reduced during the 19th century, especially following the independence of Brazil, its largest colony.

After the 1910 revolution deposed the monarchy, the unstable Portuguese First Republic was established being then superseded by the "Estado Novo" dictatorship. Democracy was restored after the Portuguese Colonial War and the Carnation Revolution in 1974, after which Portugal's last overseas provinces became independent (most prominently Angola and Mozambique); the last overseas territory, Macau, was ceded to China in 1999.

Portugal is a developed country with an advanced and high-income economy, with a very high Human Development Index. It has the world's 19th-highest quality-of-life, one of the top health care systems, and it's also one of the world's most globalized and peaceful nations.[11] A member of the European Union and the United Nations, it is as well a founding member of the Latin Union, the Organization of Ibero-American States, OECD, NATO, Community of Portuguese Language Countries, the Eurozone and the Schengen Agreement.

History

Early history

The Roman Temple of Diana, Évora.

The early history of Portugal is shared with the rest of the Iberian Peninsula. The name of Portugal derives from the Roman name Portus Cale. The region was settled by Pre-Celts and Celts, giving origin to peoples like the Gallaeci, Lusitanians, Celtici and Cynetes, visited by Phoenicians and Carthaginians, incorporated in the Roman Republic dominions as Lusitania and part of Gallaecia (both part of Hispania), after 45 BC until 298 AD, settled again by Suebi, Buri, and Visigoths, and conquered by Moors. Other minor influences include some 5th century vestiges of Alan settlement, which were found in Alenquer, Coimbra and even Lisbon.[12]

Reconquista

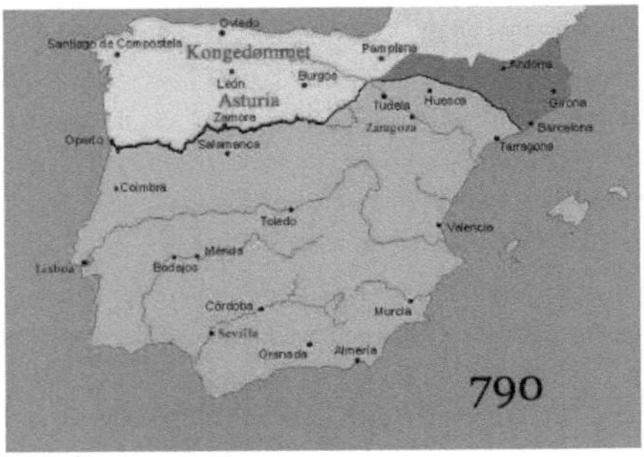

The Reconquista, 790–1300

During the Reconquista period, Christians reconquered the Iberian Peninsula from the Muslim and Moorish domination. In 868, the First County of Portugal was formed. A victory over the Muslims at Battle of Ourique in 1139 is traditionally taken as the occasion when the County of Portugal as a fief of the Kingdom of León was transformed into the independent Kingdom of Portugal.

Henry, to whom the newly formed county was awarded by Alfonso VI for his role in reconquering land from the Moors, based his newly formed county in Bracara Augusta (nowadays Braga), capital city of the ancient Roman province, and also previous capital of several kingdoms over the first millennia.

On 24 June 1128, the Battle of São Mamede occurred near Guimarães. Afonso Henriques, Count of Portugal, defeated his mother Countess Teresa and her lover Fernão Peres de Trava, thereby establishing himself as sole leader. Afonso Henriques officially declared Portugal's independence when he proclaimed himself king of Portugal on 25 July 1139, after the Battle of Ourique. He was recognized as such in 1143 by Alfonso VII, king of León and Castile, and in 1179 by Pope Alexander III.

The Castle of Guimarães, Guimarães – the city is known as the cradle of Portugal.

Afonso Henriques and his successors, aided by military monastic orders, pushed southward to drive out the Moors, as the size of Portugal covered about half of its present area. In 1249, this Reconquista ended with the capture of the Algarve on the southern coast, giving Portugal its present-day borders, with minor exceptions.

In 1348 and 1349, like the rest of Europe, Portugal was devastated by the Black Death.[13]

In 1373, Portugal made an alliance with England, which is the longest-standing alliance in the world.

In 1383, the king of Castile, husband of the daughter of the Portuguese king who had died without a male heir, claimed his throne. An ensuing popular revolt led to the 1383-1385 Crisis. A faction of petty noblemen and commoners, led by John of Aviz (later John I), seconded by General Nuno Álvares Pereira defeated the Castilians in the Battle of Aljubarrota. This celebrated battle is still a symbol of glory and the struggle for independence from neighboring Spain.

Exploration, colonization and trade

Vasco da Gama lands at Calicut (Indian Subcontinent), on May 20th 1498.

In the following decades, Portugal spearheaded the exploration of the world and undertook the Age of Discovery. Infante Dom Henry the Navigator, son of King João I, became the main sponsor and patron of this endeavor.

In 1415, Portugal conquered the first of its overseas colonies by conquering Ceuta. It was the first prosperous Islamic trade center in North Africa. There followed the first discoveries in the Atlantic: Madeira and the Azores, which led to the first colonization movements.

Throughout the 15th century, Portuguese explorers sailed the coast of Africa, establishing trading posts for several common types of tradable commodities at the time, ranging from gold to slaves, as they looked for a route to India and its spices, which were coveted in Europe.

The Treaty of Tordesillas, intended to resolve the dispute that had been created following the return of Christopher Columbus, was signed on 7 June 1494, and divided the newly discovered lands outside Europe between Portugal and Spain along a meridian 370 leagues west of the Cape Verde islands (off the west coast of Africa).

In 1498, Vasco da Gama finally reached India and brought economic prosperity to Portugal and its population of 1.7 million residents.

In 1500, Pedro Álvares Cabral discovered Brazil and claimed it for Portugal.[14] Ten years later, Afonso de Albuquerque conquered Goa, in India, Ormuz in the Persian Strait, and Malacca, now a state in Malaysia. Thus, the Portuguese empire held dominion over commerce in the Indian Ocean and South Atlantic. The Portuguese sailors set out to reach Eastern Asia by sailing eastward from Europe landing in such places as Taiwan, Japan, the island of Timor, and may have been the first Europeans to discover Australia and even New Zealand.[15]

The Treaty of Zaragoza, signed on 22 April 1529 between Portugal and Spain, specified the antimeridian to the line of demarcation specified in the Treaty of Tordesillas. All these facts made Portugal the world's major economic, military, and political power from the 15th century to the beginning of the 16th century.

Iberian Union and Restoration

Acclamation of John IV as King of Portugal. Painting by Veloso Salgado, in the Military Museum, Lisbon.

Portugal's independence was interrupted between 1580 and 1640. This occurred because the last two kings of the House of Aviz – King Sebastian, who died in the battle of Alcácer Quibir in Morocco, and his great-uncle and successor, King Henry of Portugal – both died without heirs, resulting in the extinction of that royal house. Subsequently, Philip II of Spain claimed the throne and so became Philip I of Portugal. Although Portugal did not lose its formal independence, it was governed by the same monarch who governed Spain, briefly forming a union of kingdoms, as a personal union. The joining of the two crowns deprived Portugal of a separate foreign policy, and led to the involvement in the Eighty Years' War being fought in Europe at the time between Spain and the Netherlands. War led to a deterioration of the relations with Portugal's oldest ally, England, and the loss of Hormuz. From 1595 to 1663 the Dutch-Portuguese War primarily involved the Dutch companies invading many Portuguese colonies and commercial interests in Brazil, Africa, India and the Far East, resulting in the loss of the Portuguese Indian Sea trade monopoly.

An anachronistic map of the Portuguese Empire (1415–1999). Red – actual possessions; Olive – explorations; Orange – areas of influence and trade; Pink – claims of sovereignty; Green – trading posts; Blue – main sea explorations, routes and areas of influence. The disputed Portuguese discovery of Australia is not shown.

In 1640, John IV spearheaded an uprising backed by disgruntled nobles and was proclaimed king. The Portuguese Restoration War between Portugal and Spain on the aftermath of the 1640 revolt, ended the sixty-year period of the Iberian Union under the House of Habsburg. This was the beginning of the House of Braganza, which reigned in Portugal until 1910.

According to the historian Leslie Bethell, "In 1700 Portugal had a population of about two million people. During the eighteenth century approximately 400,000 left for [the Portuguese colony of] Brazil, despite efforts by the crown to place severe restrictions on emigration."[16]

Pombaline era

Sebastião José de Carvalho e Melo, Marquis of Pombal.

In 1738, Sebastião José de Carvalho e Melo, the talented son of a Lisbon squire, began a diplomatic career as the Portuguese Ambassador in London and later in Vienna. The Queen consort of Portugal, Archduchess Maria Anne Josefa of Austria, was fond of Melo; and after his first wife died, she arranged the widowed de Melo's second marriage to the daughter of the Austrian Field Marshal Leopold Josef, Count von Daun. King John V of Portugal, however, was not pleased and recalled Melo to Portugal in 1749. John V died the following year and his son, Joseph I of Portugal was crowned. In contrast to his father, Joseph I was fond of de Melo, and with the Queen Mother's approval, he appointed Melo as Minister of Foreign Affairs. As the King's confidence in de Melo increased, the King entrusted him with more control of the state. By 1755, Sebastião de Melo was made Prime Minister. Impressed by British economic success he had witnessed while Ambassador, he successfully implemented similar economic policies in Portugal. He abolished slavery in Portugal and in the Portuguese colonies in India; reorganized the army and the navy; restructured the University of Coimbra, and ended discrimination against different Christian sects in Portugal.

But Sebastião de Melo's greatest reforms were economic and financial, with the creation of several companies and guilds to regulate every commercial activity. He demarcated the region for production of Port to ensure the wine's quality, and this was the first attempt to control wine quality and production in Europe. He ruled with a strong hand by imposing strict law upon all classes of Portuguese society from the high nobility to the poorest working class, along with a widespread review of the country's tax system. These reforms gained him enemies in the upper classes, especially among the high nobility, who despised him as a social upstart.

1755 copper engraving showing Lisbon in flames and a tsunami overwhelming the ships in the harbor after the 1755 Lisbon earthquake.

Disaster fell upon Portugal in the morning of 1 November 1755, when Lisbon was struck by a violent earthquake with an estimated Richter scale magnitude of 9. The city was razed to the ground by the earthquake and the subsequent tsunami and ensuing fires.[17] Sebastião de Melo survived by a stroke of luck and then immediately embarked on rebuilding the city, with his famous quote: "What now? We bury the dead and feed the living."

Despite the calamity and huge death toll, Lisbon suffered no epidemics and within less than one year was already being rebuilt. The new downtown of Lisbon was designed to resist subsequent earthquakes. Architectural models were built for tests, and the effects of an earthquake were simulated by marching troops around the models. The buildings and big squares of the Pombaline Downtown of Lisbon still remain as one of Lisbon's tourist attractions: They represent the world's first quake-proof buildings. Sebastião de Melo also made an important contribution to the study of seismology by designing an inquiry that was sent to every parish in the country.

Following the earthquake, Joseph I gave his Prime Minister even more power, and Sebastião de Melo became a powerful, progressive dictator. As his power grew, his enemies increased in number, and bitter disputes with the high nobility became frequent. In 1758 Joseph I was wounded in an attempted assassination. The Távora family and the Duke of Aveiro were implicated and executed after a quick trial. The Jesuits were expelled from the country and their assets confiscated by the crown. Sebastião de Melo showed no mercy and prosecuted every person involved, even women and children. This was the final stroke that broke the power of the aristocracy and ensured the victory of the Minister against his enemies. Based upon his swift resolve, Joseph I made his loyal minister Count of Oeiras in 1759.

In 1762 Spain invaded Portuguese territory as part of the Seven Years' War, however by 1763 the status-quo between Spain and Portugal of before the war had been restored.

The Portuguese royal family embarking to the Portuguese colony of Brazil following the French Invasions of Portugal; later, the Portuguese capital moved to Rio de Janeiro and the United Kingdom of Portugal, Brazil and the Algarves was established.

Following the Távora affair, the new Count of Oeiras knew no opposition. Made "Marquis of Pombal" in 1770, he effectively ruled Portugal until Joseph I's death in 1779. However, historians also argue that Pombal's "enlightenment," while far-reaching, was primarily a mechanism for enhancing autocracy at the expense of individual liberty and especially an apparatus for crushing opposition, suppressing criticism, and furthering colonial economic exploitation as well as intensifying book censorship and consolidating personal control and profit.[18]

The new ruler, Queen Maria I of Portugal, disliked the Marquis because of the power he amassed, and never forgave him for the ruthlessness at which he dispatched the Távora family, and upon her accession to the throne, she did what she had long vowed to do: she withdrew all his political offices. Pombal died peacefully on his estate at Pombal in 1782.

In the autumn of 1807, Napoleon moved French troops through Spain to invade Portugal. From 1807 to 1811, British-Portuguese forces would successfully fight against the French invasion of Portugal, while the royal family and the Portuguese nobility, including Maria I, relocated to the Portuguese territory of Brazil, at that time a colony of the Portuguese Empire, in South America. This episode is known as the Transfer of the Portuguese Court to Brazil.

Brazilian independence

Brazil just before independence in 1822 – Cisplatina (modern Uruguay) was one of the last additions to the territory of Brazil under Portuguese rule.

Portugal began a slow but inexorable decline until the 20th century. This decline was hastened by the independence in 1822 of the country's largest colonial possession, Brazil. In 1807, as Napoleon Bonaparte's army closed in on Portugal's capital city of Lisbon, the Prince Regent João VI of Portugal transferred his court to Brazil and established Rio de Janeiro as the capital of the Portuguese Empire. In 1815, the Portuguese Empire changed its name to the United Kingdom of Portugal, Brazil and the Algarves.

Due to the change in its status and the arrival of the Portuguese royal family, Brazilian administrative, civic, economical, military, educational, and scientific apparatus were expanded and highly modernized. Portuguese and their allied British troops fought against the French Invasion of Portugal and by 1815 the situation in Europe had cooled down sufficiently that João VI would be able to safely return to Lisbon. However, the King of Portugal remained in Brazil until the Liberal Revolution of 1820, which started in Porto, demanded his return to Lisbon in 1821.

Thus he returned to Portugal but left his son Pedro in charge of Brazil. When the king attempted the following year to return the Kingdom of Brazil to subordinate status as a principality, his son Pedro, with the overwhelming support of the Brazilian elites, declared Brazil's independence from Portugal. Cisplatina (today's sovereign state of Uruguay), in the south, was one of the last additions to the territory of Brazil under Portuguese rule.

Colonial Portuguese Africa

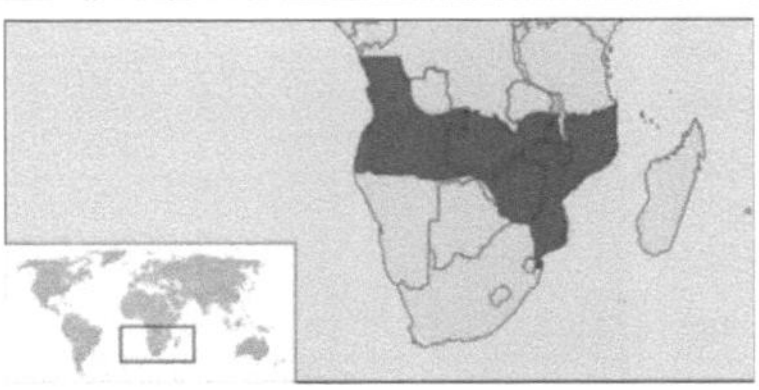

The *Pink Map* project: Portugal's claim of sovereingty over the lands between Portuguese Angola and Portuguese Mozambique.

At the height of European colonialism in the 19th century, Portugal had already lost its territory in South America and all but a few bases in Asia. Luanda, Benguela, Bissau, Lourenço Marques, Porto Amboim and the Island of Mozambique were among the oldest Portuguese-founded port cities in its African territories. During this phase, Portuguese colonialism focused on expanding its outposts in Africa into nation-sized territories to compete with other European powers there.

With the Conference of Berlin of 1884, Portuguese Africa territories had their borders formally established on request of Portugal in order to protect the centuries-long Portuguese interests in the continent from rivalries enticed by the Scramble for Africa. Portuguese Africa's cities and towns like Nova Lisboa, Sá da Bandeira, Silva Porto, Malanje, Tete, Vila Junqueiro, Vila Pery and Vila Cabral were founded or redeveloped inland during this period and beyond. New coastal towns like Beira, Moçâmedes, Lobito, João Belo, Nacala and Porto Amélia, were also founded. Even before the turn of the century, railway tracks as the Benguela railway in Angola, and the Beira railway in Mozambique, started to be built to link coastal areas and selected inland regions.

Other episodes during this period of the Portuguese presence in Africa include the 1890 British Ultimatum. This forced the Portuguese military to retreat from the land between the Portuguese colonies of Mozambique and Angola (most of present-day Zimbabwe and Zambia), which had been claimed by Portugal and included in its "Pink Map," which clashed with British aspirations to create a Cape to Cairo Railway. The Portuguese territories in Africa were

Cape Verde, São Tomé and Príncipe, Portuguese Guinea, Angola, and Mozambique. The tiny fortress of São João Baptista de Ajudá on the coast of Dahomey, was also under Portuguese rule. In addition, the country still ruled the Asian territories of Portuguese India, Portuguese Timor and Macau.

Republic

On 1 February 1908, the king Carlos I of Portugal and his heir apparent, Prince Luis Filipe, were murdered in Lisbon. Under his rule, Portugal was twice declared bankrupt – on 14 June 1892, and again on 10 May 1902 – causing social turmoil, economic disturbances, protests, revolts and criticism of the monarchy. Manuel II of Portugal become the new king, but was eventually overthrown by the 5 October 1910 revolution, which abolished the regime and instated republicanism in Portugal. Political instability and economic weaknesses were fertile ground for chaos and unrest during the Portuguese First Republic, aggravated by the Portuguese military intervention in World War I, which led to a military coup d'état in 1926 and the creation of the National Dictatorship (*Ditadura Nacional*).

Manuel II, the last King of Portugal.

This in turn led to the establishment of the right-wing dictatorship of the Estado Novo under António de Oliveira Salazar in 1933. Portugal was one of only five European countries to remain neutral in World War II. From the 1940s to the 1960s, Portugal was a founding member of NATO, OECD and the European Free Trade Association (EFTA). Gradually, new economic development projects and relocation of white mainland Portuguese citizens into the overseas colonies in Africa were initiated, with Angola and Mozambique, as the largest and richest overseas territories, being the main targets of those initiatives.

End of colonialism

After India attained Independence in 1947, the residents of Dadra and Nagar Haveli, with the support of the Indian government and the help of pro-independence organisations, liberated the territories of Dadra and Nagar Haveli from Portuguese rule in 1954.[19] In 1961, São João Baptista de Ajudá's annexation by the Republic of Dahomey was the start of a process that led to the final dissolution of the centuries-old Portuguese Empire. According to the census of 1921 São João Baptista de Ajudá had 5 inhabitants and, at the moment of the ultimatum by the Dahomey Government, it had only 2 inhabitants representing Portuguese Sovereignty. Another forcible retreat from overseas territories occurred in December 1961 when Portugal refused to relinquish the territories of Goa, Daman and Diu. As a result, the Portuguese army and navy were involved in armed conflict in its colony of Portuguese India against the Indian Armed Forces. The operations resulted in the defeat of the Portuguese defensive garrison, which was forced to surrender. The outcome was the loss of the remaining Portuguese territories in the Indian subcontinent.

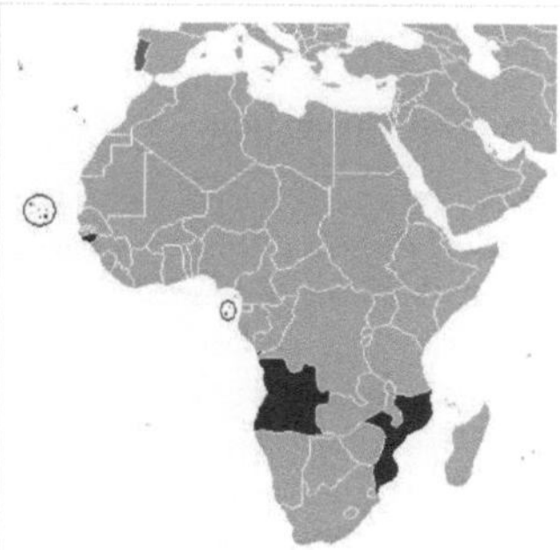

Map of the Portuguese Overseas provinces in Africa by the time of the Portuguese Colonial War (1961–1974).

Also in the early 1960s, independence movements in the Portuguese overseas provinces of Angola, Mozambique and Guinea in Africa, resulted in the Portuguese Colonial War (1961–1974).

Carnation Revolution

Throughout the colonial war period Portugal had to deal with increasing dissent, arms embargoes and other punitive sanctions imposed by most of the international community. However, the authoritarian and conservative Estado Novo regime, firstly installed and governed by António de Oliveira Salazar and from 1968 onwards led by Marcelo Caetano, tried to preserve a vast centuries-long intercontinental empire with a total area of 2,168,071 km^2.[20] The Portuguese government and army successfully resisted the decolonization of its overseas territories until April 1974, when a bloodless left-wing military coup in Lisbon, known as the Carnation Revolution, led the way for the independence of the overseas territories in Africa and Asia, as well as for the restoration of democracy after two years of a transitional period known as PREC (*Processo Revolucionário Em Curso*, or On-Going Revolutionary Process). This period was characterized by social turmoil and power disputes between left- and right-wing political forces. Some factions, including Álvaro Cunhal's *Partido Comunista Português* (PCP), unsuccessfully tried to turn the country into a communist state. The retreat from the overseas territories and the acceptance of its independence terms by Portuguese head representatives for overseas negotiations, which would create newly-independent communist states in 1975 (most notably the People's Republic of Angola and the People's Republic of Mozambique), prompted a mass exodus of Portuguese citizens from Portugal's African territories (mostly from Portuguese Angola and Mozambique).[21] [22]

Over a million destitute Portuguese refugees fled the former Portuguese colonies. Mário Soares and António de Almeida Santos were charged with organising the independence of Portugal's overseas territories. By 1975, all the Portuguese African territories were independent and Portugal held its first democratic elections in 50 years. However, the country continued to be governed by a military-civilian provisional administration until the Portuguese legislative election of 1976 that took place on 25 April, exactly one year after the previous election, and two years after the Carnation Revolution. With a new Constitution approved, the country's main aim was economic recovery and strengthening of the nation's democracy. It was won by the Portuguese Socialist Party (PS) and Mário Soares, its leader, became Prime Minister of the 1st Constitutional Government on 23 July. In the following years, Portugal's economic situation obliged the government to pursue International Monetary Fund (IMF)-monitored stabilization programs in 1977–78 and 1983–85.

European integration

In 1986, Portugal joined the European Economic Community (EEC) that later became the European Union (EU). In the following years Portugal's economy progressed considerably as result of EEC/EU structural and cohesion funds and Portuguese companies' easier access to foreign markets. Portugal's last overseas territory, Macau, was not handed over to the People's Republic of China (PRC) until 1999, under the 1987 joint declaration that set the terms for Macau's handover from Portugal to the PRC. In 2002, the independence of East Timor (Asia) was formally recognized by Portugal, after an incomplete decolonization process that was started in 1975 because of the Carnation Revolution.

The Treaty of Lisbon was signed by the European Union member states on 13 December 2007 in the Jerónimos Monastery, Lisbon

On 26 March 1995, Portugal started to implement Schengen Area rules, eliminating border controls with other Schengen members while simultaneously strengthening border controls with non-member states. In 1996 the country was a co-founder of the Community of Portuguese Language Countries (CPLP) headquartered in Lisbon. Expo '98 took place in Portugal and in 1999 it was one of the founding countries of the euro and the Eurozone.

On 5 July 2004, José Manuel Barroso, then Prime Minister of Portugal, was nominated President of the European Commission, the most powerful office in the European Union. On 1 December 2009, the Treaty of Lisbon entered into force, after had been signed by the European Union member states on 13 December 2007 in the Jerónimos Monastery, in Lisbon, enhancing the efficiency and democratic legitimacy of the Union and improving the coherence of its action.

Economic disruption in the wake of the late-2000s financial crisis led the country to negotiate in 2011 with the IMF and the European Union, through the European Financial Stability Mechanism (EFSM) and the European Financial Stability Facility (EFSF), a loan to help the country stabilise its finances.

Geography

Mount Pico, the highest peak in Portugal, on the volcanic island of Pico in the Azores.

The blue-green lakes of the Lagoa das Sete Cidades within the Sete Cidades Massif on the island of São Miguel, Azores.

Cork oak tree in a wheat field, an iconic image of the Alentejo.

The territory of Portugal includes an area in the Iberian Peninsula (referred to as *the continent* by most Portuguese) and two archipelagos in the Atlantic Ocean: the archipelagos of Madeira and the Azores. It lies between latitudes 32° and 43° N, and longitudes 32° and 6° W.

Mainland Portugal is split by its main river, the Tagus that flows from Spain and disgorges in Tagus Estuary, before escaping into the Atlantic. The northern landscape is mountainous towards the interior with several plateaus indented by river valleys, whereas the south, that includes the Algarve and the Alentejo regions, is characterized by rolling plains.

Portugal's highest peak is the similarly named Mount Pico on the island of Pico in the Azores. This ancient volcano, which measures 2351 m (7713 ft) is a highly iconic symbol of the Azores, while the Serra da Estrela on the mainland (the summit being 1991 m (6532 ft) above sea level) is an important seasonal attraction for skiers and winter sports enthusiasts.

The archipelagos of Madeira and the Azores are scattered within the Atlantic Ocean: the Azores straddling the Mid-Atlantic Ridge on a tectonic triple junction, and Madeira along a range formed by in-plate hotspot geology (much like the Hawaiian Islands). Geologically, these islands were formed by volcanic and seismic events, although the last terrestrial volcanic eruption occurred in 1957–58 (Capelinhos) and minor earthquakes occur sporadically, usually of low intensity.

Portugal's Exclusive Economic Zone, a sea zone over which the Portuguese have special rights over the exploration and use of marine resources, has 1,727,408 km^2. This is the 3rd largest Exclusive Economic Zone of the European Union and the 11th largest in the world.

Climate

Portugal is defined as a Mediterranean climate (*Csa* in the south and Douro region, and *Csb* in the north and western Alentejo), and also Semi-arid climate or Steppe climate (*Bsk* in certain parts of Beja district) according to the Koppen-Geiger Climate Classification), and is one of the warmest European countries: the annual average temperature in mainland Portugal varies from 13 °C (55.4 °F) in the mountainous interior north to over 18 °C (64.4 °F) in the south and on the Guadiana river basin. The Algarve, separated from the Alentejo region by mountains reaching up to 900 metres in Pico da Foia, has a climate similar to that of the southern coastal areas of Spain.

Praia Dona Ana in Lagos, a coastal beach typical of the Algarve.

Annual average rainfall in the mainland varies from just over 3000 mm (118.1 in) in the northern mountains to less than 300 mm (11.8 in) in the area of the Massueime River, near Côa, along the Douro river. Mount Pico is recognized as receiving the largest annual rainfall (over6250 mm (246.1 in) per year) in Portugal, according to *Instituto de Meteorologia* (English: *Portuguese Meteorological Institute*).[23]

A view of Serra da Estrela, the highest range on the mainland.

In some areas, such as the Guadiana basin, annual average temperatures can be as high as 20 °C (68 °F), but summer temperatures may be over 45 °C (113 °F) (as a study from the Archeological Park in Côa determined).[24] In the high mountains, such as Peneda-Gerês National Park, a temperate maritime climate permeates (*Cfb*, according to Koppen-Geiger). The record high of 47.4 °C (117.3 °F) was recorded in Amareleja (although this is not the hottest spot in summer, according to satellite readings).[25]

Snowfalls occur regularly in the interior North and Center of the country in particular in the districts of Vila Real, Bragança, Viseu and Guarda. In winter temperatures may drop below –10 °C (14.0 °F) in particular in Serra da Estrela, Serra do Gerês and Serra de Montesinho. In these places snow can fall any time from October to May. In the south of the country snowfalls are rare but still occur in the highest elevations.

The terraced hills of the wine-making region of the Douro Valley.

The country has around 2500 to 3200 hours of sunshine a year, an average of 4–6 h in winter and 10–12 h in the summer, with higher values in the southeast and lower in the northwest. The sea surface temperature on the west coast of mainland Portugal varies from 13 °C (55.4 °F)-15 °C (59.0 °F) in winter to 18 °C (64.4 °F)-20 °C (68.0 °F) in the summer while on the south coast in the summer the water temperature is typically around 22 °C (71.6 °F) occasionally reaching 25 °C (77.0 °F).

Both the archipelagos of the Azores and Madeira have a subtropical climate, although variations between islands exist, making weather predictions very difficult (owing to topography, temperature and humidity). The Madeiran and Azorean archipelagos have a narrower temperature range, with annual average temperatures exceeding 20 °C (68 °F) along the coast (according to the Portuguese Meteorological Institute). Some islands in Azores do have drier months in the summer. Consequently, the island of the Azores have been identified as having a Mediterranean climate (both *Csa* and *Csb* types), while some islands (such as Flores or Corvo) are classified as Maritime Temperate (*Cfb*) or Humid subtropical (*Cfa*), respectively, according to Koppen-Geiger classification. Porto Santo island in Madeira has a semi-arid Steppe climate (*BSh*). The Savage Islands, which are part of the regional territory of Madeira are unique in being classified as a Desert climates (*BWh*) with an annual average rainfall of approximately 150 mm (5.9 in). The sea surface temperature in the archipelagos varies from 16 °C (60.8 °F)-18 °C (64.4 °F) in winter to 23 °C (73.4 °F)-25 °C (77.0 °F) in the summer occasionally reaching 26 °C (78.8 °F).

In the southern Azores, and still within the Portuguese maritime territory, there is a unique area of tropical climate (as defined by Koppen-Geiger) influenced by Gulf Stream where sea surface temperatures are over 20 °C (68 °F) even during the winter (Source AEMET).

Biodiversity

Owing to humans occupying the territory of Portugal for thousands of years, little is left of the original vegetation. Protected areas of Portugal include one national park (Portuguese: *Parque Nacional*), 12 natural parks (Portuguese: *Parque Natural*), nine natural reserves (Portuguese: *Reserva Natural*), five natural monuments (Portuguese: *Monumento Natural*), and seven protected landscapes (Portuguese: *Paisagem Protegida*), which include the Parque Nacional da Peneda-Gerês, the Parque Natural da Serra da Estrela and the Paul de Arzila. These natural environments are shaped by diverse flora, and include widespread species of pine (especially the *Pinus pinaster* and *Pinus pinea* species), the chestnut (*Castanea sativa*), the cork-oak (*Quercus suber*), the holm oak (*Quercus ilex*), the Portuguese oak (*Quercus faginea*), and eucalyptus (*Eucalyptus globulus*). All are prized for their economic value. Laurissilva is a unique type of subtropical rainforest found in few areas of Europe and the world: in the Azores, and in particular on the island of Madeira, there are large forests of endemic Laurasilva forests (the latter protected as a natural heritage preserve).

There are several species of diverse mammalian fauna, including the fox, badger, Iberian lynx, Iberian Wolf, wild goat (*Capra pyrenaica*), wild cat (*Felis silvestris*), hare, weasel, polecat, chameleon, mongoose, civet, brown bear (spotted near Rio Minho, close to Peneda-Gerês) and many others. Portugal is an important stopover for migratory birds, in places such as Cape St. Vincent or the Monchique mountain, where thousands of birds cross from Europe to Africa during the autumn or in the spring (return migration). Most of the avian species congregate along the Iberian Peninsula since it is the closest stopover between northern Europe and Africa. Six hundred bird species make their nests in Portugal (either permanently or during the course of migration), and annually there are new registries of nesting species. The archipelagos of the Azores and Madeira are transient stopover for American, European, and African birds, while continental Portugal mostly encounters European and African bird species.

There are over 100 varieties of freshwater fish species, varying from the giant European catfish (in the Tagus International Natural Park) to some small and endemic species that live only in small lakes (along the western lakes for example). Some of these rare and specific species are highly endangered because of habitat loss, pollution and drought. Upwelling along the west coast of Portugal makes the sea extremely rich in nutrients and diverse species of marine fish; the Portuguese marine waters are one of the richest in the world. Marine fish species are more common, and include thousands of species, such as the sardine (*Sardina pilchardus*), tuna and Atlantic mackerel. Bioluminescent species are also well-represented (including species in different colour spectrum and forms), like the glowing plankton that are possible to observe in some beaches.

There are many endemic insect species, most only found in certain parts of Portugal, while other species are more widespread like the stag beetle (*Lucanus cervus*) and the cicada. The Macronesian islands (Azores and Madeira) have many endemic species (like birds, reptiles, bats, insects, snails and slugs) that evolved independent from other regions of Portugal. In Madeira, for example, it is possible to observe more than 250 species of land gastropods.

Governance

Portugal has been a democratic republic since the ratification of the Constitution of 1976, with Lisbon, the nation's largest city, as its capital. The constitution grants the division, or separation, of powers among legislative, executive, and judicial branches. The four main institutions as described in this constitution are the President of the Republic, the Parliament, known as the *Assembleia da República* (English: *Assembly of the Republic*), the Government, headed by a Prime Minister, and the courts.

Aníbal Cavaco Silva is the current President of the Portuguese Republic.

The President, who is elected to a five-year term, has a supervisory non-executive role: the current President is Aníbal Cavaco Silva. The Parliament is a chamber composed of 230 deputies elected for a four-year term. The government, whose head is the Prime Minister (currently Pedro Passos Coelho), chooses a *Council of Ministers*, that comprises the Ministers and State Secretaries. The courts are organized into several levels: judicial, administrative, and fiscal branches. The Supreme Courts are institutions of last resort/appeal. A thirteen-member Constitutional Court oversees the constitutionality of the laws.

Portugal operates a multi-party system of competitive legislatures/local administrative governments at the national-, regional- and local-levels. The Legislative Assembly, Regional Assemblies and local municipalities and/or parishes, are dominated by two political parties, the Socialist Party and the Social Democratic Party, in addition to the Unitarian Democratic Coalition (Portuguese Communist Party plus Ecologist Party "The Greens"), the Left Bloc and the Democratic and Social Centre – People's Party, which garner between 5 and 15% of the vote regularly.

Pedro Passos Coelho is the current Prime Minister of Portugal.

Executive branch

The President, elected to a five-year term by direct, universal suffrage, is also Commander-in-Chief of the Armed Forces. Presidential powers include the appointment of the Prime Minister and Council of Ministers (where the President is obligated by the results from Legislative Elections); dismissing the Prime Minister; dissolving the Assembly (to call early elections); vetoing legislation (which may be overridden by the Assembly); and declaring a State of War or siege.

The President is advised on issues of importance by the Council of State, which is composed of six senior civilian officers, any former Presidents elected under the 1976 Constitution, five-members chosen by the Assembly, and five selected by the president.

The Government is headed by the presidentially-appointed Prime Minister, who names a Council of Ministers to act as the government and cabinet. Each government is required to define the broad outline of its policies in a program, and present it to the Assembly for a mandatory period of debate. The failure of the Assembly to reject the program by a majority of deputies confirms the government in office.

Legislative branch

The Assembly of the Republic is a unicameral body composed of up to 230 deputies. Elected by universal suffrage according to a system of proportional representation, deputies serve four-year terms of office, unless the President dissolves the Assembly and calls for new elections.

Law and criminal justice

The Portuguese legal system is part of the civil law legal system, also called the continental family legal system. Until the end of the 19th century, French law was the main influence. Since then, the major influence has been German law. The main laws include the Constitution (1976, as amended), the Civil Code (1966, as amended) and the Penal Code (1982, as amended). Other relevant laws are the *Commercial Code* (1888, as amended) and the *Civil Procedure Code* (1961, as amended).

Portuguese law applied in the former colonies and territories and continues to be the major influence for those countries. Portugal's main police organizations are the *Guarda Nacional Republicana – GNR* (National Republican Guard), a gendarmerie; the *Polícia de Segurança Pública – PSP* (Public Security Police), a civilian police force who work in urban areas; and the *Polícia Judiciária – PJ* (Judicial Police), a highly specialized criminal investigation police that is overseen by the Public Ministry.

Portugal was one of the first countries in the world to abolish the death penalty. Maximum jail sentences are limited to 25 years.

Portugal has arguably the most liberal laws concerning possession of illicit drugs in the Western world. In 2001 Portugal decriminalized possession of effectively all drugs that are still illegal in other developed nations including, but not limited to, marijuana, cocaine, heroin, and LSD. While possession is legal, trafficking and possession of more than "10 days worth of personal use" are still punishable by jail time and fines. People caught with small amounts of any drug are given the choice to go to a rehab facility, and may refuse treatment without consequences. Despite criticism from other European nations, who stated Portugal's drug consumption would tremendously increase, overall drug use rose only slightly, whilst use among teenagers dropped, along with the number of HIV infection cases, which had dropped 50% by 2009.[26] [27]

On 31 May 2010, Portugal became the sixth country in Europe and the eighth country in the world to legally recognize same-sex marriage on the national level. The law came into force on 5 June 2010.[28]

Administrative divisions

Administratively, Portugal is divided into 308 municipalities (Portuguese: *municípios or concelhos*), which are subdivided into 4260 civil parishes (Portuguese: *freguesia*). Operationally, the municipality and civil parish, along with the national government, are the only legally identifiable local administrative units identified by the government of Portugal (for example, cities, towns or villages have no standing in law, although may be used as catchment for the defining services). For statistical purposes the Portuguese government also identifies NUTS, inter-municipal communities and informally, the district system, used until European integration (and being phased-out by the national government). Continental Portugal is agglomerated into 18 districts, while the archipelagos of the Azores and Madeira are governed as autonomous regions; the largest units, established since 1976, are either mainland Portugal (Portuguese: *Portugal Continental*) and the autonomous regions of Portugal (Azores and Madeira).

The 18 districts of mainland Portugal are: Aveiro, Beja, Braga, Bragança, Castelo Branco, Coimbra, Évora, Faro, Guarda, Leiria, Lisbon, Portalegre, Porto, Santarém, Setúbal, Viana do Castelo, Vila Real and Viseu – each district takes the name of the district capital.

Within the European Union NUTS (Nomenclature of Territorial Units for Statistics) system, Portugal is divided into seven regions: the Azores, Alentejo, Algarve, Centro, Lisboa, Madeira and Norte, and with the exception of the Azores and Madeira, these NUTS areas are subdivided into 28 subregions.

Districts[29]								
	District	Area	Population			District	Area	Population
1	Lisbon	2,761 km^2	2,244,984		10	Guarda	5,518 km^2	160,931
2	Leiria	3517 km^2 (1358 sq mi)	470,765		11	Coimbra	3,947 km^2	429,714
3	Santarém	6747 km^2 (2605 sq mi)	454,456		12	Aveiro	2,808 km^2	714,351
4	Setúbal	5064 km^2 (1955 sq mi)	849,842		13	Viseu	5,007 km^2	378,166
5	Beja	10225 km^2 (3948 sq mi)	152,706		14	Bragança	6,608 km^2	136,459
6	Faro	4960 km^2 (1915 sq mi)	450,484		15	Vila Real	4,328 km^2	207,184
7	Évora	7,393 km^2	167,434		16	Porto	2,395 km^2	1,816,045
8	Portalegre	6,065 km^2	118,952		17	Braga	2,673 km^2	848,444
9	Castelo Branco	6,675 km^2	195,949		18	Viana do Castelo	2,255 km^2	244,947

Autonomous Regions			
Autonomous Region	**Area**	**Population**	**Demonym**
Azores	2,333 km^2	246,102	*Azorean*
Madeira	801 km^2	267,938	*Madeiran*

Foreign relations

A member state of the United Nations since 1955, Portugal is also a founding member of NATO (1949), OECD (1961) and EFTA (1960); it left the latter in 1986 to join the European Economic Community, that would become the European Union in 1993. In 1996 it co-founded the Community of Portuguese Language Countries (CPLP), which seeks to foster closer economic and cultural ties between the world's Lusophone nations.

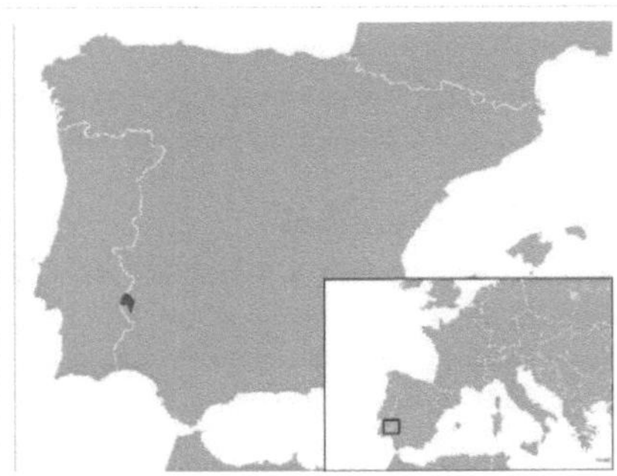

The disputed territory of Olivença

In addition, Portugal is a full member of the Latin Union (1983) and the Organization of Ibero-American States (1949). It has a friendship alliance and dual citizenship treaty with its former colony, Brazil. Portugal and England (subsequently, the United Kingdom of Great Britain and Northern Ireland) share the world's oldest active military accord through their Anglo-Portuguese Alliance (Treaty of Windsor), which was signed in 1373.

The only international dispute concerns the municipality of Olivença (Olivenza). Under Portuguese sovereignty since 1297, the municipality of Olivenza was ceded to Spain under the Treaty of Badajoz in 1801, after the War of the Oranges. Portugal claimed it back in 1815 under the Treaty of Vienna. However, since the 19th century, it has been continuously and peacefully ruled by Spain which considers the territory not only de facto but also de jure as an integral part of Spain.

Military

The armed forces have three branches: Navy, Army and Air Force. They serve primarily as a self-defense force whose mission is to protect the territorial integrity of the country and providing humanitarian assistance and security at home and abroad. As of 2008, the three branches numbered 39,000 active personnel including 6,500 women. Portuguese military expenditure in 2009 was $5.2 billion, representing 2.1 percent of GDP. Military conscription was abolished in 2004. The minimum age for voluntary recruitment is 18 years.

A Portuguese F-16 refueling in mid flight.

The Army (21,000 personnel) comprises three brigades and other small units. An infantry brigade (mainly equipped with Pandur II APC), a mechanized brigade (mainly equipped with Leopard 2 A6 tanks and M113 APC) and a Rapid Reaction Brigade (consisting of paratroopers, commandos and rangers). The Navy (10,700 personnel, of which 1,580 are marines) has five frigates, two submarines, and 28 patrol and auxiliary vessels. The Air Force (7,500 personnel) has the Lockheed F-16 Fighting Falcon as the main combat aircraft.

The *NRP Corte Real*, (Vasco da Gama class frigate) of the Portuguese Navy.

In addition to the three branches of the armed forces, there is the Republican National Guard, a security force subject to military law and organization (gendarmerie) comprising 25,000 personnel. This force is under the authority of both the Defense and the Interior Ministry. It has provided detachments for participation in international operations in Iraq and East Timor.

The United States maintains a military presence with 770 troops in the Lajes Air Base at Terceira Island, in the Azores. The Allied Joint Force Command Lisbon (JFC Lisbon) – one of the three main subdivisions of NATO's Allied Command Operations – it is based in Oeiras, near Lisbon.

In the 20th century, Portugal engaged in two major military interventions: World War I and the Portuguese Colonial War (1961–1974). After the end of the Portuguese Empire in 1975, the Portuguese Armed Forces have participated in peacekeeping missions in East Timor, Bosnia, Kosovo, Afghanistan, Somalia, Iraq (Nasiriyah) and Lebanon. Portugal also conducted several independent unilateral military operations abroad, as were the cases of the interventions of the Portuguese Armed Forces in Angola in 1992 and in Guinea-Bissau in 1998 with the main objectives of protecting and withdrawing of Portuguese and foreign citizens threatened by local civil conflits.

Economy

The municipality of Maia, in Porto Metropolitan Area, is among the most industrialized places in Portugal. Sonae, Portugal's largest private employer, was founded in Maia and has its headquarters there.

Since the Carnation Revolution (1974) which culminated with the end of one of its most notable phases of economic expansion (that started in the 1960s),[30] there has been a significant change in annual economic growth. After the turmoil of the 1974 revolution and the PREC period, Portugal has been trying to adapt itself to a changing modern global economy. Since the 1990s, Portugal's economic development model has been slowly changing from one based on public consumption to one focused on exports, private investment, and development of the high-tech sector. Business services have overtaken more traditional industries such as textiles, clothing, footwear, cork (of which Portugal is the world's leading producer),[31] wood products and beverages.[32]

Most industry, business and finance are concentrated in Lisbon and Porto metropolitan areas. The districts of Aveiro, Braga, Coimbra, and Leiria are the biggest economic centres outside those two main metropolitan areas.

The Portuguese currency is the euro (€) and the country's economy is in the Eurozone since its starting. Portugal's central bank is the *Banco de Portugal*, which is an integral part of the European System of Central Banks.

Sectors

Primary sector

Most of Portugal's farms have a small area devoted to a diversified intensive farming, like these in the Oeste subregion.

Agriculture in Portugal is based on small to medium-sized family-owned dispersed units. However, the sector also includes larger scale intensive farming export-oriented agrobusinesses backed by companies (like Grupo RAR's Vitacress, Sovena, Lactogal, Vale da Rosa, Companhia das Lezírias and Valouro). The country produces a wide variety of crops and livestock products, including green vegetables, rice, corn, barley, olives, oilseeds, nuts, cherries, bilberry, table grapes, edible mushrooms, dairy products, poultry and beef. Forestry has also played an important economic role among the rural communities and industry (namely paper industry that includes Portucel Soporcel Group, engineered wood that includes Sonae Indústria, and furniture that includes several manufacturing plants in and around Paços de Ferreira, the core of Portugal's major industrial operations of IKEA). In 2001, the gross agricultural product accounted for 4% of the national GDP.

Traditionally a sea-power, Portugal has had a strong tradition in the Portuguese fishing sector and is one of the countries with the highest fish consumption per capita.[33] The main landing sites in Portugal (including Azores and Madeira), according to total landings in weight by year, are the harbours of Matosinhos, Peniche, Olhão, Sesimbra, Figueira da Foz, Sines, Portimão and Madeira. Portuguese processed fish products are exported through several companies under a number of different brands like Conservas Ramirez, the World's oldest canned fish producer still in operation, as well as Combate, Comur, General, Líder, Maná, Murtosa, Pescador, Pitéu, Tenório, Torreira, Vasco da Gama, etc.

Secondary sector

Industry is diversified, ranging from automobile (Volkswagen Autoeuropa, Peugeot Citroen), aerospace (Embraer), electronics and textiles, to food, chemicals, cement and wood pulp. Volkswagen Group's AutoEuropa motor vehicle assembly plant in Palmela is among the largest foreign direct investment projects in Portugal.

Modern non-traditional technology-based industries like aerospace, biotechnology and information technology, have been developed in several locations across the country. Alverca, Covilhã,[34] Évora,[35] and Ponte de Sor are the main centres of Portuguese aerospace industry, which is led by Brazil-based company Embraer and the Portuguese company OGMA. Since after the turn of the 21st century, many major biotechnology and information technology industries have been founded and are concentrated in the metropolitan areas of Lisbon, Porto, Braga, Coimbra and Aveiro.

Tertiary sector

Travel and tourism continues to become extremely important for Portugal, with visitor numbers forecast to increase significantly over the next years. However, there is increasing competition from Eastern European destinations such as Croatia who offer similar attractions, which are often cheaper. Consequently, the country is almost obligated to focus on its niche attractions such as health, nature and rural tourism in order to stay ahead of its competitors.[36]

Marina beach in Vilamoura, Algarve.

The banking and insurance sectors performed well until the late-2000s financial crisis, partly reflecting a rapid deepening of the market in Portugal. While sensitive to various types of market and underwriting risks, both the life and non-life sectors, overall, are estimated to be able to withstand a number of severe shocks, even though the impact on individual insurers varies widely.[37]

State-owned companies

Major State-owned companies include Águas de Portugal (water), ANA (airports), Caixa Geral de Depósitos (banking), Comboios de Portugal (railways), Companhia das Lezírias (agriculture), CTT (postal services), RTP (media) and TAP Portugal (airline). Some of the former are managed by state-run holding company Parpública, which is a shareholder of several companies, both public and private.

Listed companies

Companies listed on Euronext Lisbon stock exchange like EDP, Cimpor, Corticeira Amorim, Galp, Jerónimo Martins, Millennium bcp, Portucel Soporcel, Portugal Telecom and Sonae, are among the largest corporations of Portugal by number of employees, net income or international market share. The Euronext Lisbon is the major stock exchange of Portugal and is part of the NYSE Euronext, the first global stock exchange. The PSI-20 is Portugal's most selective and widely known stock index.

Performance

The Global Competitiveness Report for 2005, published by the World Economic Forum, placed Portugal's competitiveness in the 22nd position, but the 2008–2009 edition placed Portugal in the 43rd position out of 134 countries and territories.[38] Research about quality of life by the Economist Intelligence Unit's quality of life survey placed Portugal as the country with the 19th-best quality of life in the world for 2005, ahead of other economically and technologically advanced countries like France, Germany, the United Kingdom and South Korea, but 9 places behind its only neighbour, Spain.[39] This is despite the fact that Portugal remains the country with the lowest per capita GDP in Western Europe.[40]

Oeiras Municipality, in Lisbon Metropolitan Area, is home of many of the headquarters of multinational companies operating in Portugal

The poor performance of the Portuguese economy was explored in April 2007 by The Economist, which described Portugal as "a new sick man of Europe".[41] From 2002 to 2007, the unemployment rate increased by 65% (270,500 unemployed citizens in 2002, 448,600 unemployed citizens in 2007).[42] By early December 2009, unemployment had reached 10.2% – a 23-year record high. In December 2009, ratings agency Standard and Poor's lowered its long-term credit assessment of Portugal to "negative" from "stable," voicing pessimism on the country's structural weaknesses in the economy and weak competitiveness that would hamper growth and the capacity to strengthen its public finances and reduce debt.[43] In July 2011, ratings agency Moody's downgraded its long-term credit assessment of Portugal after warning of deteriorating risk of default in March 2011.[44]

Corruption has become an issue of major political and economic significance for the country. Some cases are well known and were widely reported in the media, such as the affairs in several municipalities involving local town hall officials and businesspersons, as well as a number of politicians with wider responsibilities and power.[45] [46] Nevertheless the Transparency International report for 2010 places Portugal in 31st position in terms of perceived corruption, just below Israel and Spain, and 34 positions above Italy.[47]

A report published in January 2011 by the Diário de Notícias, a leading Portuguese newspaper, demonstrated that in the period between the Carnation Revolution in 1974 and 2010, the democratic Portuguese Republic governments encouraged over expenditure and investment bubbles through unclear public-private partnerships. This funded numerous ineffective and unnecessary external consultancy and advising committees and firms, allowed considerable slippage in state-managed public works, inflated top management and head officers' bonuses and wages, causing a persistent and lasting recruitment policy that boosted the number of redundant public servants. The economy was also damaged by risky credit, public debt creation and mismanaged European structural and cohesion funds for almost four decades. Apparently, the Prime Minister Sócrates's cabinet was not able to forecast or prevent any of this when symptoms first appeared in 2005, and in 2011 the country was on the verge of bankruptcy.[48]

If analysed under a wider time span, the convergence of the Portuguese economy to EU levels has been impressive, especially from 1986 to the early 2000s.[49] [50] According to Barry (2003), "what appears to have been crucial in the Portuguese case, relative to Spain at least, is the degree of labour-market flexibility that the economy exhibits. (...) Thus Portuguese convergence has been impressive, even though, consistent with its relatively low human-capital stock, the economy has specialised in low-tech production."[50]

On April 6, 2011 Prime Minister José Sócrates announced on national television that the country would request financial assistance from the IMF and the European Financial Stability Facility, like Greece and the Republic of Ireland had done before. It was the third time that external financial aid was requested to the IMF – the first was in the late 1970s following the Carnation Revolution.

In October 2011, Moody's Analytics downgraded nine Portuguese banks, blaming financial weakness.[51]

Labour market

Although a developed country and a high income country, Portugal has the lowest GDP per capita in Western Europe and its population has one of the lowest incomes per head among member states of the European Union. According to Eurostat, in 2009, Portugal's GPD per capita stood at 80% of the EU27 average,[52] the 10th lowest in the Union.

The average wage in Portugal is 1,039 € per month (net),[53] and the minimum wage, which is regulated by law, is €485 per month (although paid in 14 installments, which means that on average the minimum wage is 565 euros). Officially, in 2008 the unemployment rate decreased to 7.3% in the second quarter of 2008.[54] However, it immediately rose again to higher rates. Influenced by events worldwide, by December 2009, unemployment had surpassed the 10% mark nationwide, by 2010, it was about 11%, and in 2011 it was above 12%.

Tourism

The Pena National Palace in Sintra is a UNESCO World Heritage Site and one of Portugal's most visited landmarks

Portugal is among the 20 most visited countries in the world, receiving an average of 13 million foreign tourists each year.[55] Tourism is playing an increasingly important role in Portugal's economy, contributing to about 5% of its Gross Domestic Product (GDP).

Tourist hotspots in Portugal are Lisbon, the Algarve and Madeira, but the Portuguese government continues to promote and develop new tourist destinations, such as the Douro Valley, the island of Porto Santo, and Alentejo. Lisbon is, after Barcelona, the European city which attracts the most tourists (with seven million tourists occupying the city's hotels in 2006, a number that grew 11.8% compared to previous year).[56] Lisbon in recent years surpassed the Algarve as the leading tourist region in Portugal. Porto and Northern Portugal, especially the urban areas north of Douro River valley, was the tourist destination which grew most (11.9%) in 2006, surpassing Madeira (in 2010), as the third most visited destination.

Most tourists in Portugal are British-, Spanish- or German-origin visitors, travel by low cost airliners, and not only seek sun and beaches, but increasingly search for cultural, gastronomic, environmental or nautical experiences (or travel for reasons of business).

Tourist regions

The main tourist regions can be broken-down into (by order of importance): the Greater Lisbon (Portuguese: *Lisboa*), the Algarve, Greater Porto and Northern Portugal (Portuguese: *Porto and Norte*), the Portuguese Islands (Portuguese: *Ilhas Portuguesas*: Madeira and Azores), and Alentejo. Other tourist regions include *Douro Sul*, *Templários*, *Dão-Lafões*, *Costa do Sol*, *Costa Azul*, *Planície Dourada*, that are unknown to many tourists or visitors.

Almourol Castle, a Knights Templar stronghold used during the Christian Reconquista of Iberia from the Muslim Moors

Most of these regions are grouped in tourism reference areas, which continue to be in a state of reorganization and evolution, some based on the traditional regions of Portugal: the *Costa Verde* (*Green Coast*); *Costa da Prata* (*Silver Coast*)); *Costa de Lisboa* (*Lisbon Coast*); *Montanhas* (*Mountains*); *Planícies* (*Plains*); *Algarve*; and the islands of the archipelagos of Madeira and the Azores.

Transport

Porto Metro light rail.

The Orient Station (*Gare do Oriente*), Lisbon.

By the early 1970s Portugal's fast economic growth with increasing consumption and purchase of new automobiles set the priority for improvements in transportation. Again in the 1990s, after joining the European Economic Community, the country built many new motorways. Today, the country has a 68732 km (42708 mi) road network, of which almost 3000 km (1864 mi) are part of system of 44 motorways. Opened in 1944, the first motorway (which linked Lisbon to the National Stadium) was an innovative project that made Portugal among one of the first countries in the world to establish a motorway (this roadway eventually became the Lisbon-Cascais highway, or A5). But, although a few other tracts were created (around 1960 and 1970), it was only after the beginning of the 1980s that large-scale motorway construction was implemented. In 1972, Brisa, the highway concessionaire, was founded to handle the management of many of the regions motorways.

Vasco da Gama bridge is the longest bridge in Europe.[59] [60]

Vasco da Gama Bridge, over the Tagus River, is the longest bridge in Europe.[57] [58]

Continental Portugal's 89015 km^2 (34369 sq mi) territory is serviced by three international airports located near the principal cities of Lisbon, Porto and Faro. Lisbon's geographical position makes it a stopover for many foreign airlines at several airports within the country. The primary flag-carrier is TAP Portugal, although many other domestic airlines provide services within and without the country. The government decided to build a new airport outside Lisbon, in Alcochete, to replace Lisbon Portela Airport. Currently, the most important airports are in Lisbon, Faro, Porto, Funchal (Madeira), and Ponta Delgada (Azores), managed by the national airport authority group ANA – Aeroportos de Portugal.

25 April Bridge across the Tagus River.

A national railway system that extends throughout the continent and into Spain, is supported and administered by Comboios de Portugal. Rail transport of passengers and goods is derived using the 2791 km (1734 mi) of railway lines currently in service, of which 1430 km (889 mi) are electrified and about 900 km (559 mi) allow train speeds greater than 120 km/h (75 mph). The railway network is managed by the REFER while the transport of passengers and goods are the responsibility of Caminhos de Ferro Portugueses (CP), both public companies. In 2006 the CP carried 133 million passengers and 9750000 t (long tons; short tons) of goods.

The major seaports are located in Leixões, Aveiro, Figueira da Foz, Lisbon, Setúbal, Sines and Faro.

The two largest metropolitan areas have subway systems: Lisbon Metro and *Metro Sul do Tejo* in the Lisbon Metropolitan Area and Porto Metro in the Porto Metropolitan Area, each with more than 35 km (22 mi) of lines. In Portugal, Lisbon tram services have been supplied by the *Companhia de Carris de Ferro de Lisboa* (Carris), for over a century. In Porto, a tram network, of which only a tourist line on the shores of the Douro remain, began construction on 12 September 1895 (a first for the Iberian Peninsula). All major cities and towns have their own local urban transport network, as well as taxi services.

Science and technology

The headquarters (rectorate) of the University of Porto (UP) – UP is the largest Portuguese university by both number of students and research output.

Scientific and technological research activities in Portugal are mainly conducted within a network of R&D units belonging to public universities and state-managed autonomous research institutions like the INETI – Instituto Nacional de Engenharia, Tecnologia e Inovação and the INRB – Instituto Nacional dos Recursos Biológicos. The funding and management of this research system is mainly conducted under the authority of the Ministry of Science, Technology and Higher Education (MCTES) itself and the MCTES's Fundação para a Ciência e Tecnologia (FCT). The largest R&D units of the public universities by volume of research grants and peer-reviewed publications, include biosciences research institutions like the Instituto de Medicina Molecular, the Centre for Neuroscience and Cell Biology, the IPATIMUP, the Instituto de Biologia Molecular e Celular and the Abel Salazar Biomedical Sciences Institute.

António Egas Moniz was a Portuguese neurologist awarded with the Nobel Prize in Physiology or Medicine in 1949.

Among the largest non-state-run research institutions in Portugal are the Instituto Gulbenkian de Ciência and the Champalimaud Foundation, a neuroscience and oncology research centre, which in addition awards every year one of the highest monetary prizes of any science prize in the world. A number of both national and multinational high-tech and industrial companies, are also responsible for research and development projects. One of the oldest learned societies of Portugal is the Sciences Academy of Lisbon, founded in 1779. Iberian bilateral state-supported research efforts include the International Iberian Nanotechnology Laboratory and the Ibercivis distributed computing platform, which are joint research programmes of both Portugal and Spain. Portugal is a member of several pan-European scientific organizations. These include the European Space Agency (ESA), the European Laboratory for Particle Physics (CERN), ITER, and the European Southern Observatory (ESO).

Portugal has the largest aquarium in Europe, the Lisbon Oceanarium, and the Portuguese have several other notable organizations focused on science-related exhibits and divulgation, like the state agency *Ciência Viva*, a programme of the Portuguese Ministry of Science and Technology to the promotion of a scientific and technological culture among the Portuguese population,[61] the Science Museum of the University of Coimbra, the National Museum of Natural History at the University of Lisbon, and the Visionarium.

The Lisbon Oceanarium

With the emergence and growth of several science parks throughout the world that helped create many thousands of scientific, technological and knowledge-based businesses, Portugal started to develop several[62] science parks across the country. These include the Taguspark (in Oeiras), the Coimbra iParque (in Coimbra), the biocant (in Cantanhede), the Madeira Tecnopolo[63] (in Funchal), Sines Tecnopolo[64] (in Sines), Tecmaia[65] (in Maia) and Parkurbis[66] (in Covilhã). Companies locate in the Portuguese science parks to take advantage of a variety of services ranging from financial and legal advice through to marketing and technological support.

Egas Moniz, a Portuguese physician who developed the cerebral angiography and leucotomy, received in 1949 the Nobel Prize in Physiology or Medicine – he is the first Portuguese recipient of a Nobel Prize and the only in the sciences.

The European Innovation Scoreboard 2011, placed Portugal-based innovation in the 15th position, with an impressive increase in innovation expenditure and output.[67]

Energy

Portugal has considerable resources of wind and river power, the two most cost-effective renewable sources. Since the 2000s, there has been a trend towards the development of a renewable resource industry and reduction of both consumption and use of fossil fuel resources. In 2006, the world's largest solar power plant at that date began operating near Moura, in the south, while the world's first commercial wave power farm opened in the Norte region (October 2006). By the end of 2006, 66% of the country's electrical production was from coal and fuel power plants, while 29% were derived from hydroelectric dams, and 6% by wind energy.[68]

Alqueva Dam, in the Alentejo; the Alqueva project, an irrigation and hydroelectric power generation system, created the largest artificial lake in Western Europe.

Serpa solar power plant in the Alentejo. Around 70% of Portugal's energy production comes from renewable sources.

In 2008, renewable energy resource methods began to produce 43% of the nation's consumption of electricity, even as hydroelectric production decreased due to severe droughts.[69] As of June 2010, electricity exports have outnumbered imports; 70% of the national production of energy has transitioned to renewable sources.[70]

Portugal's national energy transmission company, Redes Energéticas Nacionais (REN), uses sophisticated modeling to predict weather, especially wind patterns, and computer programs to calculate energy from the various renewable-energy plants. Before the solar/wind revolution, Portugal had generated electricity from hydropower plants on its rivers for decades. But new programs combine wind and water: wind-driven turbines pump water uphill at night, the most blustery period; then the water flows downhill by day, generating electricity, when consumer demand is highest. Portugal's distribution system is also now a two-way street. Instead of just delivering electricity, it draws electricity from even the smallest generators, like rooftop solar panels. The government aggressively encouraged such contributions by setting a premium price for those who buy rooftop-generated solar electricity.

Portugal is a nuclear free zone by choice. Projects for the construction of nuclear power plants were always dismissed by the governments and a majority of the local populations.[71] [72]

Demographics

The *Instituto Nacional de Estatística* (Portuguese: *National Institute of Statistics*) estimates that, according to the 2011 census, the population was 10,555,853 (of which 52% was female, 48% was male). This population has been relatively homogeneous for most of its history: a single religion (Catholicism) and a single language have contributed to this ethnic and national unity, namely after the expulsion of the Moors, Moriscos and Sephardi Jews.[73]

Native Portuguese are an Iberian ethnic group, whose ancestry is very similar to other Western and Southern Europeans and Mediterranean peoples, in particular Spaniards, with whom they share a common ancestry, history and cultural proximity.

The most important demographic influence in the modern Portuguese seems to be the oldest one; current interpretation of Y-chromosome and mtDNA data suggests that the Portuguese have their origin in Paleolithic peoples that began arriving to the European continent around 45,000 years ago. All subsequent migrations did leave an impact, genetically and culturally, but the main population source of the Portuguese is still Paleolithic.

Urbanization

There are two Greater Metropolitan Areas (GAMs): Lisbon and Porto.[74]

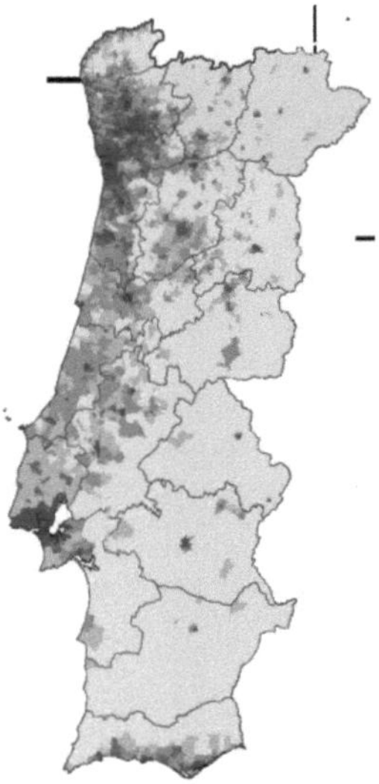

Population density map of Portugal

Rank	City name	Population	Metro Area	Population[75]	Subregion	Population
1	Lisbon	**564,657**	**Lisbon**	**2,661,850**	Grande Lisboa	**2,003,580**
2	Porto	**263,131**	**Porto**	**1,679,854**	Grande Porto	**1,572,176**
3	Braga	**181,819**	**Minho**	**814,083**	Cávado	**410,608**
4	Vila Nova de Gaia	178,255	Porto	–	Grande Porto	–
5	Amadora	175,872	Lisbon	–	Grande Lisboa	–
6	Almada	101,500	Lisbon	–	Península de Setúbal	–
7	Coimbra	**101,069**	**Coimbra**	**435,900**	Baixo Mondego	**340,342**
8	Funchal	100,526	n/a[76]	n/a	Madeira	245,806
9	Setúbal	89,303	Lisbon	–	Península de Setúbal	714,589
10	Agualva-Cacém	81,845	Lisbon	–	Grande Lisboa	–

Immigration

Portugal's colonial history has long since been a cornerstone of its national identity, as has its geographic position at the southwestern corner of Europe, looking out into the Atlantic Ocean. It was the last western colonial European powers to give up its overseas territories (among them Angola and Mozambique in 1975), turning over the administration of Macau to the People's Republic of China at the end of 1999. Consequently, it has both influenced and been influenced by cultures from former colonies or dependencies, resulting in immigration from these former territories for both economic and/or personal reasons. Portugal, long a country of emigration (the vast majority of

Brazilians have some Portuguese ancestry),[77] has now become a country of net immigration,[78] and not just from the last Indian (Portuguese until 1961), African (Portuguese until 1975), and Far East Asian (Portuguese until 1999) overseas territories. An estimated 800,000 Portuguese returned to Portugal as the country's African possessions gained independence in 1975.[77] By 2007, Portugal had 10,617,575 inhabitants of whom about 332,137 were legal immigrants.[1]

Since the 1990s, along with a boom in construction, several new waves of Ukrainian, Brazilian, people from the former Portuguese colonies in Africa and other Africans have settled in the country. Romanians, Moldovans and Chinese have also chosen Portugal as destination. Portugal's Romani population, estimated at about 40,000,[79] offers another element of ethnic diversity. Most Romanis congregate with similar ethnic groups in the southern parts of the country and sell clothing and handicrafts in rural markets.

In addition, a number of EU citizens, mostly from the United Kingdom, northern European or Nordic countries, have become permanent residents in the country (with the British community being mostly composed of retired pensioners and choosing to live in the Algarve and Madeira).[80]

Religion

Today, 84.5% of the Portuguese population are Roman Catholic while 2.2% follow other Christian faiths.[81] Some 9% of the population are self-declared as non-religious (Zuckerman 2005). In addition, the country has small Protestant, Mormon, Muslim, Hindu, Sikh, Christian Orthodox, Jehovah's Witnesses, Baha'i, Buddhist and Jewish communities.

The Sanctuary of Our Lady of Fátima, the largest Marian shrine.

Many Portuguese holidays, festivals and traditions have a Christian origin or connotation. Although relations between the Portuguese state and the Roman Catholic Church were generally amiable and stable since the earliest years of the Portuguese nation, their relative power fluctuated. In the 13th and 14th centuries, the church enjoyed both riches and power stemming from its role in the reconquest, its close identification with early Portuguese nationalism and the foundation of the Portuguese educational system, including the first university. The growth of the Portuguese overseas empire made its missionaries important agents of colonization, with important roles in the education and evangelization of people from all the inhabited continents. The growth of liberal and nascent republican movements during the eras leading to the formation of the First Portuguese Republic (1910–26) changed the role and importance of organized religion.

Portugal is a secular state: church and state were formally separated during the Portuguese First Republic, and later reiterated in the 1976 Portuguese Constitution. Other than the Constitution, the two most important documents relating to religious freedom in Portugal are: the 1940 Concordata (later amended in 1971) between Portugal and the Holy See, and the 2001 Religious Freedom Act.

Languages

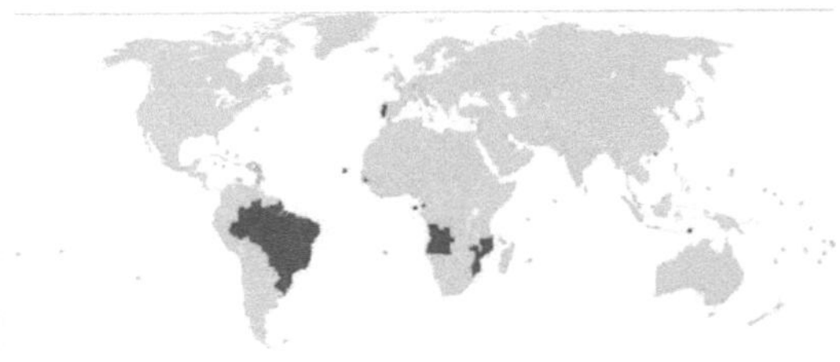

Map of the Lusosphere

Portuguese is the official language of Portugal. Portuguese is a Romance language that originated in what is now Galicia (Spain) and Northern Portugal, from the Galician-Portuguese language. It is derived from the Latin spoken by the romanized Pre-Roman peoples of the Iberian Peninsula around 2000 years ago. In the 15th and 16th centuries, it spread worldwide as Portugal established a colonial and commercial empire (1415–1999).

As a result, nowadays the Portuguese language is also official and spoken in Brazil, Angola, Mozambique, Cape Verde, São Tomé and Príncipe, Guinea-Bissau, and East Timor. These countries, plus Macau Special Administrative Region (People's Republic of China), make up the Lusosphere, term derived from the ancient Roman province of *Lusitania*, which currently matches the Portuguese territory south of the Douro river. Mirandese is also recognized as a co-official regional language in some municipalities of northeastern Portugal. It retains fewer than 5,000 speakers in Portugal (a number that can be up to 12,000 if counting second language speakers).

Education

Headquarters of the New University of Lisbon.

The educational system is divided into preschool (for those under age 6), basic education (9 years, in three stages, compulsory), secondary education (3 years, till the 12th grade), and higher education (university and polytechnic).

Total adult literacy rate is 99%. Portuguese primary school enrollments are close to 100%. According to the OECD's Programme for International Student Assessment (PISA) 2009, the average Portuguese 15-years old student, when rated in terms of reading literacy, mathematics and science knowledge, is placed at the same level as those students from the United States, Sweden, Germany, Ireland, France, Denmark, United Kingdom, Hungary and Taipei, with 489 points (493 is the average).[82] Over 35% of college-age citizens (20 years old) attend one of the country's higher education institutions[83] (compared with 50% in the United States and 35% in the OECD countries). In addition to being a key destination for international students, Portugal is also among the top places of origin for international students. All higher education students, both domestic and international, totaled 380,937 in 2005.

The tower of the University of Coimbra, Coimbra – the university is one of the oldest in continuous operation in the world.

Portuguese universities have existed since 1290. The oldest Portuguese university was first established in Lisbon before moving to Coimbra. Historically, within the scope of the Portuguese Empire, the Portuguese founded in 1792 the oldest engineering school of Latin America (the Real Academia de Artilharia, Fortificação e Desenho), as well as the oldest medical college of Asia (the Escola Médico-Cirúrgica de Goa) in 1842. The largest university in Portugal is the University of Porto. Universities are usually organized into faculties. Institutes and schools are also common designations for autonomous subdivisions of Portuguese higher education institutions. The Bologna process has been adopted since 2006 by Portuguese universities and polytechnical institutes. Higher education in state-run educational establishments is provided on a competitive basis, a system of *numerus clausus* is enforced through a national database on student admissions. However, every higher education institution offers also a number of additional vacant places through other extraordinary admission processes for sportsmen, mature applicants (over 23 years old), international students, foreign students from the Lusosphere, degree owners from other institutions, students from other institutions (academic transfer), former students (readmission), and course change, which are subject to specific standards and regulations set by each institution or course department. Most student costs are supported with public money. However, with the increasing tuition fees a student has to pay to attend a Portuguese state-run higher education institution and the attraction of new types of students (many as part time students or in evening classes) like employees, businessmen, parents, and pensioners, many departments make a substantial profit from every additional student enrolled in courses, with benefits for the college or university's gross tuition revenue and without loss of educational quality (teacher per student, computer per student, classroom size per student, etc).

Portugal has entered into cooperation agreements with MIT (US) and other North American institutions to further develop and increase the effectiveness of Portuguese higher education and research.

Health

According to the latest Human Development Report, the average Life Expectancy in 2010 was 79.1 years.

The Portuguese health system is characterized by three coexisting systems: the National Health Service (NHS), special social health insurance schemes for certain professions (health subsystems) and voluntary private health insurance. The NHS provides universal coverage. In addition, about 25% of the population is covered by the health subsystems, 10% by private insurance schemes and another 7% by mutual funds.

Hospital of Santa Maria, Lisbon.

The Ministry of Health is responsible for developing health policy as well as managing the NHS. Five regional health administrations are in charge of implementing the national health policy objectives, developing guidelines and protocols and supervising health care delivery. Decentralization efforts have aimed at shifting financial and management responsibility to the regional level. In practice, however, the autonomy of regional health administrations over budget setting and spending has been limited to primary care.

The NHS is predominantly funded through general taxation. Employer (including the state) and employee contributions represent the main funding sources of the health subsystems. In addition, direct payments by the patient and voluntary health insurance premiums account for a large proportion of funding.

Similar to the other Eur-A countries, most Portuguese die from noncommunicable diseases. Mortality from cardiovascular diseases (CVD) is higher than in the Eurozone, but its two main components, ischaemic heart disease and cerebrovascular disease, display inverse trends compared with the Eur-A, with cerebrovascular disease being the single biggest killer in Portugal (17%). Portuguese people die 12% less often from cancer than in the Eur-A, but mortality is not declining as rapidly as in the Eur-A. Cancer is more frequent among children as well as among women younger than 44 years. Although lung cancer (slowly increasing among women) and breast cancer (decreasing rapidly) are scarcer, cancer of the cervix and the prostate are more frequent. Portugal has the highest mortality rate for diabetes in the Eur-A, with a sharp increase since the late 1980s.

Portugal's infant mortality rate has dropped sharply since the 1980s, when 24 of 1000 newborns died in the first year of life. It is now around 3 deaths per a 1000 newborns. This improvement was mainly due to the decrease in neonatal mortality, from 15.5 to 3.4 per 1000 live births.

People are usually well informed about their health status, the positive and negative effects of their behaviour on their health and their use of health care services. Yet their perceptions of their health can differ from what administrative and examination-based data show about levels of illness within populations. Thus, survey results based on self-reporting at the household level complement other data on health status and the use of services. Only one third of adults rated their health as good or very good in Portugal (Kasmel et al., 2004). This is the lowest of the Eur-A countries reporting and reflects the relatively adverse situation of the country in terms of mortality and selected morbidity.[84]

Culture

Portugal has developed a specific culture while being influenced by various civilizations that have crossed the Mediterranean and the European continent, or were introduced when it played an active role during the Age of Discovery. In the 1990s and 2000s, Portugal modernized its public cultural facilities, in addition to the Calouste Gulbenkian Foundation established in 1956 in Lisbon. These include the Belém Cultural Center in Lisbon, Serralves Foundation and the Casa da Música, both in Porto, as well as new public cultural facilities like municipal libraries and concert halls that were built or renovated in many municipalities across the country.

Belém Tower, Lisbon. A UNESCO World Heritage Site and a typical example of Portugal's unique Manueline architecture.

Architecture

Traditional architecture is distinctive and include the Manueline, also known as Portuguese late Gothic, a sumptuous, composite Portuguese style of architectural ornamentation of the first decades of the 16th century, incorporating maritime elements and representations of the Portuguese Age of Discovery. Modern Portugal has given the world renowned architects like Eduardo Souto de Moura, Álvaro Siza Vieira (both Pritzker Prize winners) and Gonçalo Byrne. In Portugal Tomás Taveira is also noteworthy, particularly due to stadium design.[85] [86] [87]

Cinema

Portuguese cinema has a long tradition, reaching back to the birth of the medium in the late 19th century. Portuguese film directors such as Arthur Duarte, António Lopes Ribeiro, Pedro Costa, Manoel de Oliveira, António-Pedro Vasconcelos, João César Monteiro, João Botelho and Leonel Vieira, are among those that gained notability. Noted Portuguese film actors include Joaquim de Almeida, Daniela Ruah, Maria de Medeiros, Diogo Infante, Soraia Chaves, Vasco Santana, Ribeirinho, and António Silva, among many others

Literature

Luís de Camões, Portuguese poet of the 16th century.

Portuguese literature, one of the earliest Western literatures, developed through text and song. Until 1350, the Portuguese-Galician troubadours spread their literary influence to most of the Iberian Peninsula.[88] Gil Vicente (ca. 1465 – ca. 1536), was one of the founders of both Portuguese and Spanish dramatic traditions.

Adventurer and poet Luís de Camões (ca. 1524–1580) wrote the epic poem *"Os Lusíadas"* (The Lusiads), with Virgil's Aeneid as his main influence. Modern Portuguese poetry is rooted in neoclassic and contemporary styles, as exemplified by Fernando Pessoa (1888–1935). Modern Portuguese literature is represented by authors such as Almeida Garrett, Camilo Castelo Branco, Eça de Queiroz, Sophia de Mello Breyner Andresen, António Lobo Antunes and Miguel Torga. Particularly popular and distinguished is José Saramago, winner of the 1998 Nobel Prize for literature.

Cuisine

Bacalhau à Brás.

Portuguese cuisine is diverse. The Portuguese consume a lot of dry cod (*bacalhau* in Portuguese), for which there are hundreds of recipes. There are more than enough *bacalhau* dishes for each day of the year. Two other popular fish recipes are grilled sardines and caldeirada, a potato-based stew that can be made from several types of fish. Typical Portuguese meat recipes, that may be made out of beef, pork, lamb, or chicken, include *cozido à portuguesa*, feijoada, frango de churrasco, *leitão* (roast suckling pig) and *carne de porco à alentejana*, a very popular northern dish is the arroz de sarrabulho (rice stewed in pigs blood) or the arroz de cabidela (Rice and chickens meat stewed in chickens blood).

Typical fast food dishes include the francesinha from Porto, and *bifanas* (grilled pork) or *prego* (grilled beef) sandwiches, which are well known around the country. The Portuguese art of pastry has its origins in Middle-Ages Catholic monasteries widely spread across the country. These monasteries, using very few ingredients (mostly almonds, flour, eggs and some liquor), managed to create a spectacular wide range of different pastries, of which pastéis de Belém (or *pastéis de nata*) originally from Lisbon, and ovos moles from Aveiro are examples. Portuguese cuisine is very diverse, with different regions having their own traditional dishes. The Portuguese have a culture of good food and throughout the country there are myriad good restaurants and small typical *tascas*.

Portuguese wines have deserved international recognition since the times of the Roman Empire, which associated Portugal with their god Bacchus. Today the country is known by wine lovers and its wines have won several international prizes. Some of the best Portuguese wines are: Vinho Verde, Vinho Alvarinho, Vinho do Douro, Vinho do Alentejo, Vinho do Dão, Vinho da Bairrada and the sweet: Port Wine, Madeira Wine and the Moscatel from Setúbal and Favaios. Port Wine is well known around the world and the most widely known wine type in the world. The Douro wine region is the oldest in the world.

Music

The Fado, painting by Portuguese artist José Malhoa (1855–1933).

Portuguese music encompasses a wide variety of genres. The most renowned is fado, a melancholy urban music, usually associated with the Portuguese guitar and *saudade*, or longing. Coimbra fado, a unique type of fado, is also noteworthy. Internationally notable performers include Amália Rodrigues, Carlos Paredes, José Afonso, Mariza, Carlos do Carmo, António Chainho, Mísia, and Madredeus. One of the most notable Portuguese musical groups outside the country, and especially in Germany, is the goth-metal band Moonspell.

In addition to fado and folk, the Portuguese listen to pop and other types of modern music, particularly from North America and the United Kingdom, as well as a wide range of Portuguese and Brazilian artists and bands. Bands with international recognition include Blasted Mechanism and The Gift, both of which were nominated for an MTV Europe Music Award.

Casa da Música, a concert hall in Porto.

Portugal has several summer music festivals, such as *Festival Sudoeste* in Zambujeira do Mar, *Festival de Paredes de Coura* in Paredes de Coura, *Festival Vilar de Mouros* near Caminha, and *Optimus Alive!*, *Rock in Rio Lisboa* and *Super Bock Super Rock* in Greater Lisbon. Out of the summer season, Portugal has a large number of festivals, designed more to an urban audience, like Flowfest or Hip Hop Porto. Furthermore, one of the largest international Goa trance festivals takes place in central Portugal every two years, and the student festivals of *Queima das Fitas* are major events in a number of cities across Portugal. In 2005, Portugal held the MTV Europe Music Awards, in Pavilhão Atlântico, Lisbon.

Fandango is one of the most popular regional dances.

In the Classical music domain, Portugal is represented by names as the pianist Artur Pizarro, Maria João Pires, Sequeira Costa, the violinists Gerardo Ribeiro, Carlos Damas, and in the past by the great cellist Guilhermina Suggia. Notable composers include José Vianna da Motta, Carlos Seixas, João Domingos Bomtempo, João de Sousa Carvalho, Luís de Freitas Branco and his student Joly Braga Santos, Fernando Lopes-Graça, Emmanuel Nunes and Sérgio Azevedo.

Painting

It has also a rich history as far as painting is concerned. The first well-known painters date back to the 15th century – like Nuno Gonçalves – were part of the Gothic painting period. José Malhoa, known for his work *Fado*, and Columbano Bordalo Pinheiro (who painted the portraits of Teófilo Braga and Antero de Quental) were both references in naturalist painting.

The 20th century saw the arrival of Modernism, and along with it came the most prominent Portuguese painters: Amadeo de Souza-Cardoso, who was heavily influenced by French painters, particularly by the Delaunays. Among his best known works is *Canção Popular a Russa e o Fígaro*. Another great modernist painter/writer was Almada Negreiros, friend to the poet Fernando Pessoa, who painted his (Pessoa's) portrait. He was deeply influenced by both Cubist and Futurist trends. Prominent international figures in visual arts nowadays include painters Vieira da Silva, Júlio Pomar, Helena Almeida, Joana Vasconcelos, Julião Sarmento and Paula Rego.

Sport

Football (soccer) is the most popular and played sport. There are several football competitions ranging from local amateur to world-class professional level. The legendary Eusébio is still a major symbol of Portuguese football history. FIFA World Player of the Year winners Luís Figo and Cristiano Ronaldo, are among the numerous examples of other world-class football (soccer) players born in Portugal and noted worldwide. Portuguese football managers are also noteworthy, with José Mourinho, André Villas-Boas, Carlos Queiroz and Manuel José among the most renowned.

Opening ceremony of the UEFA Euro 2004, at Estádio do Dragão, Porto.

The Portuguese national teams, have titles in the FIFA World Youth Championship and in the UEFA youth championships. The main national team – *Selecção Nacional* – finished second in Euro 2004 (held in Portugal), reached the third place in the 1966 FIFA World Cup, and reached the fourth place in the 2006 FIFA World Cup, their best results in major competitions to date.

Sport Lisboa e Benfica, Futebol Clube do Porto, and Sporting Clube de Portugal are the largest sports clubs by popularity and by number of trophies won, often known as "*os três grandes*" ("the big three"). They have 12 titles won in the European UEFA club competitions, were present in many finals and have been regular contenders in the last stages almost every season. Other than football, many Portuguese sports clubs, including the "big three", compete in several other sports events with a varying level of success and popularity, these may include basketball, futsal, handball, and volleyball.

Portugal has a successful rink hockey team, with 15 world titles and 20 European titles, making it the country with the most wins in both competitions. The most successful Portuguese rink hockey clubs in the history of European championships are Futebol Clube do Porto, Sporting Clube de Portugal, Sport Lisboa e Benfica and Óquei de Barcelos.

The national rugby union team made a dramatic qualification into the 2007 Rugby World Cup and became the first all amateur team to qualify for the World Cup since the dawn of the professional era. The Portuguese national rugby sevens team has performed well, becoming one of the strongest teams in Europe, and proved their status as European champions in several occasions.

Pavilhão Atlântico (Atlantic Pavilion), an indoor sports venue and concert hall in Lisbon.

In athletics, the Portuguese have won a number of gold, silver and bronze medals in the European, World and Olympic Games competitions. Cycling, with Volta a Portugal being the most important race, is also a popular sports event and include professional cycling teams such as Sport Lisboa e Benfica, Boavista, Clube de Ciclismo de Tavira, and União Ciclista da Maia.

The country has also achieved notable performances in sports like fencing, judo, kitesurf, rowing, sailing, surfing, shooting, triathlon and windsurf, owning several European and world titles. The paralympic athletes have also conquered many medals in sports like swimming, boccia and wrestling.

Surfing and Body Boarding are very popular in Portugal. In the picture the *World Surfing Championships* being held in Madeira.

In motor sport, Portugal is internationally noted for the Rally of Portugal, and the Estoril, Algarve Circuits and the revived Porto Street Circuit which holds a stage of the WTCC every two years, as well as for a number of internationally noted pilots in varied motor sports.

In equestrian sports, Portugal won the only Horseball-Pato World Championship (in 2006), achieved the third position in the First Horseball World Cup (organized in Ponte de Lima, Portugal, in 2008), and has achieved several victories in the European Working Equitation Championship.

In swimming sports, Portugal has two major sports: Swimming and Water Polo.

Northern Portugal has its own original martial art, *Jogo do Pau*, in which the fighters use staffs to confront one or several opponents.

Other popular sport-related recreational outdoor activities with thousands of enthusiasts nationwide include airsoft, fishing, golf, hiking, hunting and orienteering.

See also

- Portugal–Spain relations

References

Notes

[1] ITDS, Rui Campos, Pedro Senos. "Statistics Portugal" (http://www.ine.pt/xportal/xmain?xpid=INE&xpgid=ine_main). Ine.pt. . Retrieved 2011-07-02.

[2] Pordata, "Base de Dados Portugal Contemporâneo" (http://www.pordata.pt/). Accessed on March 7, 2011.

[3] http://www.ine.pt/scripts/flex_v10/Main.html

[4] "Portugal" (http://www.imf.org/external/pubs/ft/weo/2011/01/weodata/weorept.aspx?sy=2008&ey=2011&scsm=1&ssd=1&sort=country&ds=.&br=1&c=182&s=NGDPD,NGDPDPC,PPPGDP,PPPPC,LP&grp=0&a=&pr.x=40&pr.y=5). International Monetary Fund. . Retrieved 6 May 2011.

[5] "Gini Index" (http://www.ine.pt/xportal/xmain?xpid=INE&xpgid=ine_destaques&DESTAQUESdest_boui=107634627&DESTAQUESmodo=2&xlang=en). Instituto Nacional de Estatística. . Retrieved 14 July 2011.

[6] "Human Development Report 2010" (http://hdr.undp.org/en/media/HDR_2010_EN_Table1.pdf). United Nations. 2010. . Retrieved 5 November 2010.

[7] The Euromosaic study, Mirandese in Portugal (http://ec.europa.eu/education/policies/lang/languages/langmin/euromosaic/pt1_en.html#13), europa.eu – European Commission website. Retrieved January 2007.

[8] "Leite de Vasconcelos, José. *Cale e Portucale*. Opúsculos Vol. V – Etnologia (Parte I) Lisboa, Imprensa Nacional, 1938" (http://www.instituto-camoes.pt/CVC/bdc/etnologia/opusculos/vol05/opusculos05_28_38.pdf) (in Portuguese). .

[9] Brian Jenkins, Spyros A. Sofos, "Nation and identity in contemporary Europe" (http://books.google.com/books?id=LNRyNG9NNkcC&lpg=1), p.145 Routledge, 1996, ISBN 0-415-12313-5

[10] Melvin Eugene Page, Penny M. Sonnenburg, p. 481

[11] http://www.visionofhumanity.org/wp-content/uploads/2011/05/2011-GPI-Results-Report-Final.pdf

[12] Milhazes, José. Os antepassados caucasianos dos portugueses (http://www.rtp.pt/index.php?article=264957&visual=16&rss=0) – Rádio e Televisão de Portugal in Portuguese.

[13] Black Death (http://www.abc.net.au/science/articles/2007/09/13/2031252.htm?site=science/greatmomentsinscience), Great Moments in Science, ABC Science

[14] The standard view of historians is that Cabral was blown off course as he was navigating the currents of the South Atlantic, sighted the coast of South America, thereby accidentally discovering Brazil. However, for an alternative account of the discovery of Brazil, see History of Brazil

[15] Map proves Portuguese discovered Australia: new book (http://www.reuters.com/article/topNews/idUSSYD3449720070321), in Reuters (Wed 21 Mar 2007) – (see Theory of Portuguese discovery of Australia)
[16] Leslie Bethell (1986). " *The Cambridge history of Latin America: Colonial Latin America* (http://books.google.com/books?id=hhNfVshMw64C&pg=PA47&dq&hl=en#v=onepage&q=&f=false)". Cambridge University Press. p.47. ISBN 0-521-24516-8
[17] " Historical Depictions of the 1755 Lisbon Earthquake (http://nisee.berkeley.edu/lisbon/index.html)"
[18] Kenneth Maxwell, *Pombal, Paradox of the Enlightenment* (Cambridge: Cambridge University Press, 1995), 83, 91–108, 160–62.
[19] P S Lele, Dadra and Nagar Haveli: past and present, Published by Usha P. Lele, 1987,
[20] "Portugal Não É Um País Pequeno" (http://purl.pt/11440/1/P1.html). Purl.pt. . Retrieved 2011-02-21.
[21] Flight from Angola (http://www.economist.com/world/mideast-africa/displayStory.cfm?story_id=12079340), The Economist (16 August 1975).
[22] Dismantling the Portuguese Empire (http://www.time.com/time/magazine/article/0,9171,913229-1,00.html), Time Magazine (Monday, 7 July 1975).
[23] "Instituto de Meteorologia, IP Portugal" (http://www.meteo.pt). Meteo.pt. 2010-01-15. . Retrieved 27 January 2010.
[24] (http://www.ipa.min-cultura.pt/pubs/R.../folder/01.pdf)
[25] "Instituto de Meteorologia, IP Portugal" (http://www.meteo.pt/pt/oclima/extremos/). Meteo.pt. . Retrieved 22 August 2010.
[26] http://www.scientificamerican.com/article.cfm?id=portugal-drug-decriminalization
[27] Maia Szalavitz, Drugs in Portugal: Did Decriminalization Work? (http://www.time.com/time/health/article/0,8599,1893946,00.html), Time (April 26, 2009)
[28] http://dre.pt/pdf1sdip/2010/05/10500/0185301853.pdf
[29] "Districts of Portugal" (http://www.distritosdeportugal.com/). Distritosdeportugal.com. . Retrieved 22 August 2010.
[30] (Portuguese) Fundação da SEDES – As primeiras motivações (http://www.sedes.pt/conteudo.aspx?args=1,2), "Nos anos 60 e até 1973 teve lugar, provavelmente, o mais rápido período de crescimento económico da nossa História, traduzido na industrialização, na expansão do turismo, no comércio com a EFTA, no desenvolvimento dos sectores financeiros, investimento estrangeiro e grandes projectos de infra-estruturas. Em consequência, os indicadores de rendimentos e consumo acompanham essa evolução, reforçados ainda pelas remessas de emigrantes.", SEDES
[31] *Grande Enciclopédia Universal*, p. 10543, "Portugal", para. 4
[32] Investing in Portugal Report (http://www.ft.com/reports/investportugal2008), Financial Times
[33] PESSOA, M.F.; MENDES, B.; OLIVEIRA, J.S.. "Culturas marinhas em Portugal" (http://igbp-portugal.org/cluster2006/comunicacoes/m_f_pessoa/cluster2006-zc06c-comun.pdf) (in Portuguese). . "O consumo médio anual em produtos do mar pela população portuguesa, estima-se em cerca de 58,5 kg/ por habitante sendo, por isso, o maior consumidor em produtos marinhos da Europa e um dos quatro países a nível mundial com uma dieta à base de produtos do mar."
[34] (Portuguese) Aleia vai montar avião até agora vendido em kit e jactos portugueses em 2011 (http://noticias.sapo.pt/lusa/artigo/3010ee885ea3612ea6bfc1.htmlCovilhã:), 14 April 2008
[35] (Portuguese) Évora aprova isenções fiscais aos projectos da Embraer (http://diariodigital.sapo.pt/dinheiro_digital/news.asp?section_id=6&id_news=103403), Diário Digital (22 August 2008)
[36] "Travel and Tourism in Portugal" (http://www.euromonitor.com/Travel_and_Tourism_in_Portugal). Euromonitor.com. . Retrieved 2011-07-02.
[37] Portugal: Financial System Stability Assessment, including Reports on the Observance of Standards and Codes on the following topics: Banking Supervision, Securities Regulation, and Insurance Regulation (http://www.imf.org/external/pubs/ft/scr/2006/cr06378.pdf), IMF, (October 2006)
[38] "The Global Competitiveness Index rankings" (http://www.weforum.org/pdf/gcr/2008/rankings.pdf). World Economic Forum. . Retrieved 20 March 2009.
[39] *The Economist*. http://www.economist.com/media/pdf/QUALITY_OF_LIFE.pdf.
[40] (Portuguese) "Portugueses perderam poder de compra entre 2005 e 2007 e estão na cauda da Zona Euro" (http://economia.publico.pt/Noticia/portugueses-perderam-poder-de-compra-entre-2005-e-2007-e-estao-na-cauda-da-zona-euro_1352732) (in Portuguese). Público. 2008-12-11. . Retrieved 10 October 2010.
[41] "The Portuguese economy: A new sick man of Europe" (http://www.economist.com/world/europe/displaystory.cfm?story_id=9009032). The Economist. 2007-04-12. . Retrieved 2011-07-02.
[42] Luis Miguel Mota, População desempregada aumentou 65% em cinco anos (http://www.destak.pt/artigos.php?art=11796), Destak.pt (6 June 2008)
[43] Standard and Poor's pessimistic on Portugal (http://news.ph.msn.com/business/article.aspx?cp-documentid=3737721), Agence France-Presse (7 December 2009)
[44] As Portugal Default Risk Climbs, Europe Begins to Stabilize (http://www.moodysanalytics.com/~/media/Insight/Capital-Markets-Research/Market-Signals-Reviews/11-29-03-Soverign-Risk-Report.ashx), (29 March 2011)
[45] Eurojust chief embroiled in Portuguese corruption scandal (http://euobserver.com/9/28120), euobserver.com (13 May 2009)
[46] People & Power, Al Jazeera (http://q208.wordpress.com/2008/03/29/reportagem-da-al-jazeera-sobre-portugal-um-dos-paises-mais-pobres-e-corruptos-da-europa), Al Jazeera (March 2008)
[47] Transparency International report (http://www.transparency.org/policy_research/surveys_indices/cpi/2010/results), Transparency International (2010)

[48] (**Portuguese**) Grande investigação DN Conheça o verdadeiro peso do Estado (http://dn.sapo.pt/inicio/tv/interior.aspx?content_id=1750097&seccao=Media), Diário de Notícias (January 7, 2011)

[49] http://www.sciencedirect.com/science?_ob=ArticleURL&_udi=B6V97-3VWTCDS-9&_user=499884&_coverDate=11%2F30%2F1996&_rdoc=1&_fmt=high&_orig=gateway&_origin=gateway&_sort=d&_docanchor=&view=c&_acct=C000024499&_version=1&_urlVersion=0&_userid=499884&md5=d36edb1470dfd4db9b46db274f86c7a1&searchtype=a

[50] http://www.tcd.ie/business/staff/fbarry/papers/papers/Regiona_Aid_paper_revised.PDF

[51] UK financial firms downgraded by Moody's rating agency (http://www.bbc.co.uk/news/business-15211230), BBC (October 7, 2011)

[52] "Eurostat press release GDP per inhabitant 2009" (http://epp.eurostat.ec.europa.eu/cache/ITY_PUBLIC/2-15122010-BP/EN/2-15122010-BP-EN.PDF) (in Portuguese). . Retrieved 3 January 2011.

[53] "Wages and taxes for the Average Joe in the EU 27" (http://www.langlophone.com/20100526_edition/20100526_EU27_data_table_flipped.pdf). . Retrieved 14 May 2011.

[54] "Taxa de desemprego desce para 7,3 por cento no segundo trimestre" (http://economia.publico.clix.pt/noticia.aspx?id=1338969) (in Portuguese). Público. 14 August 2008. .

[55] http://www.unwto.org/facts/eng/pdf/indicators/ITA_top25.pdf

[56] DN Online, ed (25 January 2007). "Cidades atraem mais turistas do que os destinos sol e mar" (http://www.dn.pt/inicio/interior.aspx?content_id=651813) (in Portuguese). Lisbon, Portugal: Diário das Noticias. . Retrieved 30 April 2011.

[57] "ListAfterList.com" (http://www.listafterlist.com/tabid/57/listid/7961/Travel++Places/Top+10+Longest+Bridges+in+the+World.aspx). ListAfterList.com. . Retrieved 2010-08-22.

[58] "Curious? Read" (http://www.curiousread.com/2008/02/ten-longest-bridges-in-world.html). Curiousread.com. . Retrieved 2010-08-22.

[59] "ListAfterList.com" (http://www.listafterlist.com/tabid/57/listid/7961/Travel++Places/Top+10+Longest+Bridges+in+the+World.aspx). ListAfterList.com. . Retrieved 22 August 2010.

[60] "Curious? Read" (http://www.curiousread.com/2008/02/ten-longest-bridges-in-world.html). Curiousread.com. . Retrieved 22 August 2010.

[61] "Ciência Viva" (http://www.cienciaviva.pt/cienciaviva/index.asp?accao=changelang&lang=en). Cienciaviva.pt. . Retrieved 22 August 2010.

[62] "Tecparques – Associação Portuguesa de Parques de Ciência e Tecnologia" (http://www.tecparques.pt/associados.htm). Tecparques.pt. . Retrieved 22 August 2010.

[63] "Madeira Tecnopolo" (http://www.madeiratecnopolo.pt/). Madeiratecnopolo.pt. . Retrieved 22 August 2010.

[64] "Sines Tecnopolo" (http://www.sinestecnopolo.org/). Sines Tecnopolo. . Retrieved 22 August 2010.

[65] "TECMAIA Parque de Ciência e Tecnologia da Maia" (http://www.tecmaia.com.pt). Tecmaia.com.pt. . Retrieved 2011-07-02.

[66] "Parque de Ciência e Tecnologia da Covilhã (Parkurbis)" (http://www.parkurbis.pt/). Parkurbis.pt. . Retrieved 22 August 2010.

[67] Portugal ganha terreno no ranking da inovação (http://economia.publico.pt/Noticia/portugal-ganha-terreno-no-ranking-da-inovacao_1478035), Público (February 1, 2011)

[68] "IEA Energy Statistics: Portugal" (http://www.iea.org/Textbase/stats/electricitydata.asp?COUNTRY_CODE=PT). International Energy Agency. 2006. . Retrieved 7 April 2009.

[69] Staff (8 April 2009). "Fontes renováveis originaram 43% da electricidade consumida" (http://diariodigital.sapo.pt/news.asp?section_id=114&id_news=381941) (in Portuguese). *Diário Digital*. . Retrieved 17 April 2009.

[70] Staff (8 June 2010). "Portugal já exportou mais electricidade este ano que em 2009" (http://www.agenciafinanceira.iol.pt/empresas/portugal-agencia-financeira-ren-energia-electricidade/1168567-1728.html?utm_source=feedburner&utm_medium=feed&utm_campaign=Feed:+iol/agenciafinanceira+(agenciafinanceira)) (in Portuguese). *Agência Financeira*. . Retrieved 8 June 2010.

[71] Lorenzo Cimarossa, Model for Evaluation of Nuclear Energy Costs in Portugal (https://docs.google.com/viewer?a=v&q=cache:U14CiwL5PgEJ:https://dspace.ist.utl.pt/bitstream/2295/787477/1/Lorenzo%20Cimarossa%20-%20Model%20for%20Evaluation%20of%20Nuclear%20Energy%20Costs%20in%20Portugal.pdf+nuclear+energy+portugal+ferrel&hl=pt-PT&pid=bl&srcid=ADGEESgItBWJHU3zXpKWpw5EfUOtdTyk6tK9v20rV1I5W9QxPO2B7cn2z3qo6fMk6Lxi5D1MVhbSvhZ6MFDfspzc555VI1AcqcomrMi[illegible]sig=AHIEtbQcl584WCDhX_CuTHxOu2cdYSkI6g&pli=1), *Instituto Superior Técnico* - Technical University of Lisbon (December 2010); pg.12 2.5.2 History of Nuclear Energy in Portugal

[72] "Nuclear power: When the steam clears" (http://www.economist.com/node/18441163). *The Economist*. March 24, 2011. .

[73] " Portugal (http://www.jewishvirtuallibrary.org/jsource/vjw/Portugal.html)". The Virtual Jewish History Tour.

[74] Censos 2011 (http://www.ine.pt/scripts/flex_v10/Main.html), (*Instituto Nacional de Estatística*-19 July 2011)

[75] Source of the city populations: INE census, 2001.

[76] The Autonomous Region of Madeira is not a Metropolitan Area.

[77] Portugal – Emigration (http://countrystudies.us/portugal/48.htm), Eric Solsten, ed. Portugal: A Country Study. Washington: GPO for the Library of Congress, 1993.

[78] " Portugal sees integration progress (http://news.bbc.co.uk/2/hi/europe/4436276.stm)", BBC News, 14 November 2005

[79] A mais discriminada (http://criasnoticias.wordpress.com/2008/04/06/etnia-cigana-a-mais-discriminada-8-de-abril-dia-internacional-dos-ciganos/Etniacigana.), (*Expresso*-5 April 2008)

[80] Brasileiros são a maior colónia estrangeira em Portugal (http://embaixada-portugal-brasil.blogspot.com/2007/06/brasileiros-so-maior-colnia-estrangeira.html) Embaixada de Portugal No Brasil

[81] "CIA — The World Factbook – Portugal" (https://www.cia.gov/library/publications/the-world-factbook/geos/po.html). Cia.gov. . Retrieved 2011-02-21.

[82] "Alunos portugueses pela primeira vez "perto da média" – relatório PISA" (http://www.destak.pt/artigo/82223-alunos-portugueses-pela-primeira-vez-perto-da-media-relatorio-pisa) (in Portuguese). Destak. .

[83] "Um Contrato de confiança no Ensino Superior para o futuro de Portugal" (http://www.portugal.gov.pt/pt/GC18/Governo/Ministerios/MCTES/Intervencoes/Pages/20100111_MCTES_Int_Contrato_Confianca_EnsSup.aspx) (in Portuguese). Government of Portugal, portugal.gov.pt. .

[84] (http://www.euro.who.int/document/chh/por_highlights.pdf)

[85] **(Portuguese)** Estádios de Tomás Taveira e Souto Moura premiados (http://www.dn.pt/Inicio/interior.aspx?content_id=615671), Diário de Notícias (July 8, 2005)

[86] Tomás Taveira (http://www.amazon.com/TomÃ¡s-Taveira-Geoffrey-Broadbent/dp/185490034X), Geoffrey Broadbent (introduction), Publisher: St Martins Pr (February 1991)

[87] **(Portuguese)** Tomás Tveira desenha estádio do Palmeiras no Brasil (http://www.arquitectura.pt/forum/f11/sao-paulo-est-dio-do-palmeiras-no-brasil-tom-s-taveira-10811/), Diarioeconomico.com

[88] *Poesia e Prosa Medievais*, p. 9, para. 4

Sources

- Bliss, Jorge Braga de; Bliss, C.J. (1990). *Unity with Diversity in the European Economy: the Community's Southern Frontier*. London, England: Centre for Economic Policy Research. ISBN 978-0-521-39520-5.
- Juang, Noelle Anne (2008). *Africa and the Americas: Culture, Politics, and History: A Multidisciplinary Encyclopedia*. **2**. ISBN 978-1-85109-441-7.
- Page, Penny M. (2003). *Colonialism: An International, Social, Cultural, and Political Encyclopedia*. **2**. ISBN 978-1-57607-335-3.
- Brockey, Liam Matthew (2008). *Portuguese Colonial Cities in the Early Modern World*. ISBN 978-0-7546-6313-3.
- Ribeiro, Ângelo (2004) (in Portuguese). *História de Portugal I — A Formação do Território [History of Portugal: The Formation of the Territory]*. QuidNovi. ISBN 989-554-106-6.
- Ribeiro, José Hermano (2004) (in Portuguese). *História de Portugal II — A Afirmação do País [History of Portugal II - A Affirmation of Nation]*. QuidNovi. ISBN 989-554-107-4.
- de Macedo, José Hermano (2004) (in Portuguese). *História de Portugal III — A Epopeia dos Descobrimentos [History of Portugal III - The Epoch of Discoveries]*. QuidNovi. ISBN 989-554-108-2.
- de Macedo, José Hermano (2004) (in Portuguese). *História de Portugal IV — Glória e Declínio do Império [History of Portugal IV - Glory and Decline of Empire]*. QuidNovi. ISBN 989-554-109-0.
- de Macedo, José Hermano (2004) (in Portuguese). *História de Portugal V — A Restauração da Indepêndencia [History of Portugal IV - The Restauration of Independence]*. QuidNovi. ISBN 989-554-110-4.
- Saraiva, José Hermano (2004) (in Portuguese). *História de Portugal X — A Terceira República [History of Portugal X - The Third Republic]*. QuidNovi. ISBN 989-554-115-5.
- Loução, Paulo Alexandre (2000) (in Portuguese). *Portugal, Terra de Mistérios [Portugal: Land of Mysteries]* (3rd ed.). Ésquilo. ISBN 972-8605-04-8.
- Muñoz, Mauricio Pasto (2003) (in Portuguese). *Viriato, A Luta pela Liberdade [Viriato: The Struggle for Liberty]* (3rd ed.). Ésquilo. ISBN 972-8605-23-4.
- *Grande Enciclopédia Universal*. Durclub. 2004.
- (in Portuguese) *Constituição da República Portuguesa [Constitution of the Portuguese Republic]* (VI Revisão Constitucional ed.). 2004.

External links

Government

- Official Portuguese Government website (http://www.portugal.gov.pt/Portal/EN/) (English)/(Portuguese)
- Official Parliament website (http://www.en.parlamento.pt/)
- Chief of State and Cabinet Members (https://www.cia.gov/library/publications/world-leaders-1/world-leaders-p/portugal.html)

General information

- Portugal (https://www.cia.gov/library/publications/the-world-factbook/geos/po.html) entry at *The World Factbook*
- Portugal (http://ucblibraries.colorado.edu/govpubs/for/portugal.htm) at *UCB Libraries GovPubs*
- Portugal (http://www.dmoz.org/Regional/Europe/Portugal/) at the Open Directory Project
- Portugal in Photography 2007 (http://worldphotobase.com/europe/portugal.php?s=thumb)
- National English language newspaper (http://www.theportugalnews.com/)
- Wikimedia Atlas of Portugal
- National Wine Website (http://www.viniportugal.pt/)

Travel

- Official Travel and Tourism office website (http://www.visitportugal.com/Cultures/en-US/default.html)
- Official Portuguese Government Travel/media website (http://www.insideportugaltravel.com/)

kbd:Португал gag:Portugaliya koi:Португал ltg:Portugaleja rue:Портуґалія

Paleontology

Paleontology investigates the entire history of life on Earth.

Paleontology (pronounced /ˌpæliɒnˈtɒlədʒi/; British: **palaeontology**)[1] is the study of prehistoric life, including organisms' evolution and interactions with each other and their environments (their paleoecology). As a "historical science" it tries to explain causes rather than conduct experiments to observe effects. Paleontological observations have been documented as far back as the 5th century BC. The science became established in the 18th century as a result of Georges Cuvier's work on comparative anatomy, and developed rapidly in the 19th century. Fossils found in China since the 1990s have provided new information about the earliest evolution of animals, early fish, dinosaurs and the evolution of birds and mammals. Paleontology lies on the border between biology and geology, and shares with archaeology a border that is difficult to define. It now uses techniques drawn from a wide range of sciences, including biochemistry, mathematics and engineering. As knowledge has increased, paleontology has developed specialized sub-divisions, some of which focus on different types of fossil organisms while others study ecology and environmental history, such as ancient climates.

A paleontologist at work at John Day Fossil Beds National Monument

Body fossils and trace fossils are the principal types of evidence about ancient life, and geochemical evidence has helped to decipher the evolution of life before there were organisms large enough to leave fossils. Estimating the dates of these remains is essential but difficult: sometimes adjacent rock layers allow radiometric dating, which provides absolute dates that are accurate to within 0.5%, but more often paleontologists have to rely on relative dating by solving the "jigsaw puzzles" of biostratigraphy. Classifying ancient organisms is also difficult, as many do not fit well into the Linnean taxonomy that is commonly used for classifying living organisms, and paleontologists more often use cladistics to draw up evolutionary "family trees". The final quarter of the 20th century saw the development of molecular phylogenetics, which investigates how closely organisms are related by measuring how similar the DNA is in their genomes. Molecular phylogenetics has also been used to estimate the dates when species diverged, but there is controversy about the reliability of the molecular clock on which such estimates depend.

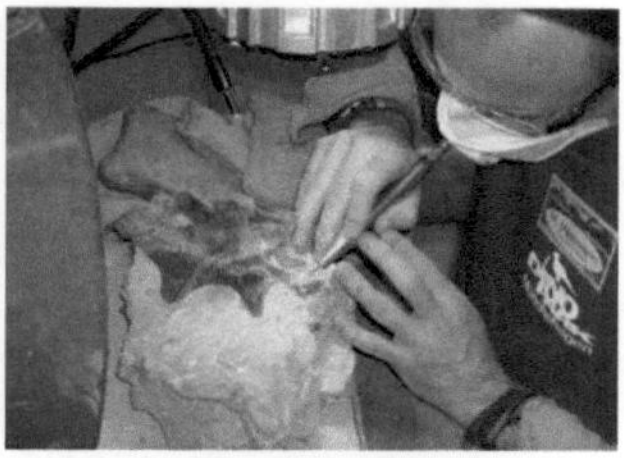

The preparation of the fossilized bones of *Europasaurus holgeri*

Use of all these techniques has enabled paleontologists to discover much of the evolutionary history of life, almost all the way back to when Earth became capable of supporting life, about 3800 [2] million years ago. For about half of that time the only life was single-celled micro-organisms, mostly in microbial mats that formed ecosystems only a few millimeters thick. Earth's atmosphere originally contained virtually no oxygen, and its oxygenation began about 2400 [3] million years ago. This may have caused an accelerating increase in the diversity and complexity of life, and early multicellular plants and fungi have been found in rocks dated from 1700 to 1200 [4] million years ago. The earliest multicellular animal fossils are much later, from about 580 [5] million years ago, but animals diversified very rapidly and there is a lively debate about whether most of this happened in a relatively short Cambrian explosion or started earlier but has been hidden by lack of fossils. All of these organisms lived in water, but plants and invertebrates started colonizing land from about 490.0 [6] million years ago and vertebrates followed them about 370 [7] million years ago. The first dinosaurs appeared about 230 [8] million years ago and birds evolved from one dinosaur group about 150 [9] million years ago. During the time of the dinosaurs, mammals' ancestors survived only as small, mainly nocturnal insectivores, but after the non-avian dinosaurs became extinct in the Cretaceous–Tertiary extinction event 65 [10] million years ago mammals diversified rapidly. Flowering plants appeared and rapidly diversified between 130 million years ago and 90 million years ago, possibly helped by coevolution with pollinating insects. Social insects appeared around the same time and, although they have relatively few species, now form over 50% of the total mass of all insects. Humans evolved from a lineage of upright-walking apes whose earliest fossils date from over 6 [11] million years ago, and anatomically modern humans appeared under 200,000 years ago. The course of evolution has been changed several times by mass extinctions that wiped out previously dominant groups and allowed other to rise from obscurity to become major components of ecosystems.

Definition

The simplest definition is "the study of ancient life".[12] Paleontology seeks information about several aspects of past organisms: "their identity and origin, their environment and evolution, and what they can tell us about the Earth's organic and inorganic past".[13]

A paleontologist carefully chips rock from a column of dinosaur vertebrae.

A historical science

Paleontology is one of the historical sciences, along with archaeology, geology, biology, astronomy, cosmology, philology and history itself.[14] This means that it aims to describe phenomena of the past and reconstruct their causes.[15] Hence it has three main elements: description of the phenomena; developing a general theory about the causes of various types of change; and applying those theories to specific facts.[14]

When trying to explain past phenomena, paleontologists and other historical scientists often construct a set of hypotheses about the causes and then look for a "smoking gun", a piece of evidence which indicates that one of the hypotheses is a better explanation than the others. Sometimes the "smoking gun" is discovered by a fortunate accident during other research, for example the discovery by Luis Alvarez and Walter Alvarez of an iridium-rich layer at the Cretaceous–Tertiary boundary made asteroid impact and volcanism the most favored explanations for the Cretaceous–Tertiary extinction event.[15]

The other main type of science is experimental science, which is often said to work by conducting experiments to *disprove* hypotheses about the workings and causes of natural phenomena – note that this approach cannot prove a hypothesis is correct, since some later experiment may disprove it. However, when confronted with totally unexpected phenomena, such as the first evidence for invisible radiation, experimental scientists often use the same approach as historical scientists: construct a set of hypotheses about the causes and then look for a "smoking gun".[15]

Related sciences

Paleontology lies on the boundary between biology and geology since paleontology focuses on the record of past life but its main source of evidence is fossils, which are found in rocks.[16] For historical reasons paleontology is part of the geology departments of many universities, because in the 19th century and early 20th century geology departments found paleontological evidence important for estimating the ages of rocks while biology departments showed little interest.[17]

Paleontology also has some overlap with archaeology, which primarily works with objects made by humans and with human remains, while paleontologists are interested in the characteristics and evolution of humans as organisms. When dealing with evidence about humans, archaeologists and paleontologists may work together – for example paleontologists might identify animal or plant fossils around an archaeological site, to discover what the people who lived there ate; or they might analyze the climate at the time when the site was inhabited by humans.[18]

In addition paleontology often uses techniques derived from other sciences, including biology, ecology, chemistry, physics and mathematics.[12] For example geochemical signatures from rocks may help to discover when life first arose on Earth,[19] and analyses of carbon isotope ratios may help to identify climate changes and even to explain major transitions such as the Permian–Triassic extinction event.[20] A relatively recent discipline, molecular phylogenetics, often helps by using comparisons of different modern organisms' DNA and RNA to re-construct evolutionary "family trees"; it has also been used to estimate the dates of important evolutionary developments, although this approach is controversial because of doubts about the reliability of the "molecular clock".[21] Techniques developed in engineering have been used to analyse how ancient organisms might have worked, for example how fast *Tyrannosaurus* could move and how powerful its bite was.[22] [23]

Analyses using engineering techniques show that *Tyrannosaurus* had a devastating bite, but raise doubts about how fast it could move.

Paleontology even contributes to astrobiology, the investigation of possible life on other planets, by developing models of how life may have arisen and by providing techniques for detecting evidence of life.[24]

Subdivisions

As knowledge has increased, paleontology has developed specialised subdivisons.[25] Vertebrate paleontology concentrates on fossils of vertebrates, from the earliest fish to the immediate ancestors of modern mammals. Invertebrate paleontology deals with fossils of invertebrates such as molluscs, arthropods, annelid worms and echinoderms. Paleobotany focuses on the study of fossil plants, but traditionally includes the study of fossil algae and fungi. Palynology, the study of pollen and spores produced by land plants and protists, straddles the border between paleontology and botany, as it deals with both living and fossil organisms. Micropaleontology deals with all microscopic fossil organisms, regardless of the group to which they belong.[26]

Instead of focusing on individual organisms, paleoecology examines the interactions between different organisms, such as their places in food chains, and the two-way interaction between organisms and their environment[27] – for example the development of oxygenic photosynthesis by bacteria hugely increased the productivity and diversity of ecosystems,[28] and also caused the oxygenation of the atmosphere, which in turn was a prerequisite for the evolution of the most complex eucaryotic cells, from which all multicellular organisms are built.[29] Paleoclimatology, although sometimes treated as part of paleoecology,[26] focuses more on the history of Earth's climate and the mechanisms which have changed it[30] – which have sometimes included evolutionary developments, for example the rapid expansion of land plants in the Devonian period removed more carbon dioxide from the atmosphere, reducing the greenhouse effect and thus helping to cause an ice age in the Carboniferous period.[31]

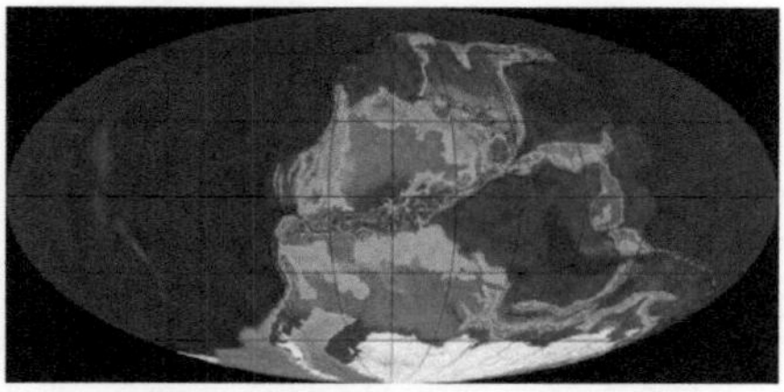

In the Carboniferous period, the continents were not in the same places as they are today, and there was extensive glaciation.

Biostratigraphy, the use of fossils to work out the chronological order in which rocks were formed, is useful to both paleontologists and geologists.[32] Biogeography studies the spatial distribution of organisms, and is also linked to geology, which explains how Earth's geography has changed over time.[33]

Sources of evidence

Body fossils

Fossils of organisms' bodies are usually the most informative type of evidence. The most common types are wood, bones, and shells.[34] Fossilisation is a rare event, and most fossils are destroyed by erosion or metamorphism before they can be observed. Hence the fossil record is very incomplete, increasingly so further back in time. Despite this, it is often adequate to illustrate the broader patterns of life's history.[35] There are also biases in the fossil record: different environments are more favorable to the preservation of different types of organism or parts of organisms.[36] Further, only the parts of organisms that were already mineralised are usually preserved, such as the shells of molluscs. Since most animal species are soft-bodied, they decay before they can become fossilised. As a result, although there are 30-plus phyla of living animals, two-thirds have never been found as fossils.[37]

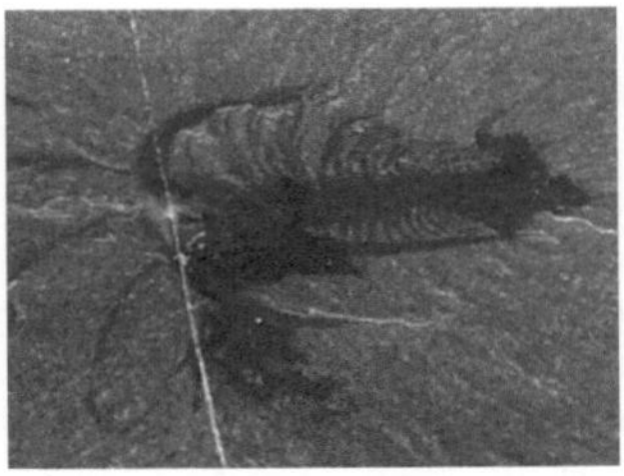

This *Marrella* specimen illustrates how clear and detailed the fossils from the Burgess Shale lagerstätte are.

Occasionally, unusual environments may preserve soft tissues. These lagerstätten allow paleontologists to examine the internal anatomy of animals that in other sediments are represented only by shells, spines, claws, etc. – if they are preserved at all. However, even lagerstätten present an incomplete picture of life at the time. The majority of organisms living at the time are probably not represented because lagerstätten are restricted to a narrow range of environments, e.g. where soft-bodied organisms can be preserved very quickly by events such as mudslides; and the exceptional events that cause quick burial make it difficult to study the normal environments of the animals.[38] The sparseness of the fossil record means that organisms are expected to exist long before and after they are found in the fossil record – this is known as the Signor-Lipps effect.[39]

Trace fossils

Trace fossils consist mainly of tracks and burrows, but also include coprolites (fossil feces) and marks left by feeding.[34] [40] Trace fossils are particularly significant because they represent a data source that is not limited to animals with easily-fossilized hard parts, and which reflects organisms' behaviour. Also many traces date from significantly earlier than the body fossils of animals that are thought to have been capable of making them.[41] Whilst exact assignment of trace fossils to their makers is generally impossible, traces may for example provide the earliest physical evidence of the appearance of moderately complex animals (comparable to earthworms).[40]

Cambrian trace fossils including *Rusophycus*, made by a trilobite

Geochemical observations

Geochemical observations may help to deduce the global level of biological activity, or the affinity of a certain fossil. For example geochemical features of rocks may reveal when life first arose on Earth,[19] and may provide evidence of the presence of eucaryotic cells, the type from which all multicellular organisms are built.[42] Analyses of carbon isotope ratios may help to explain major transitions such as the Permian–Triassic extinction event.[20]

Classifying ancient organisms

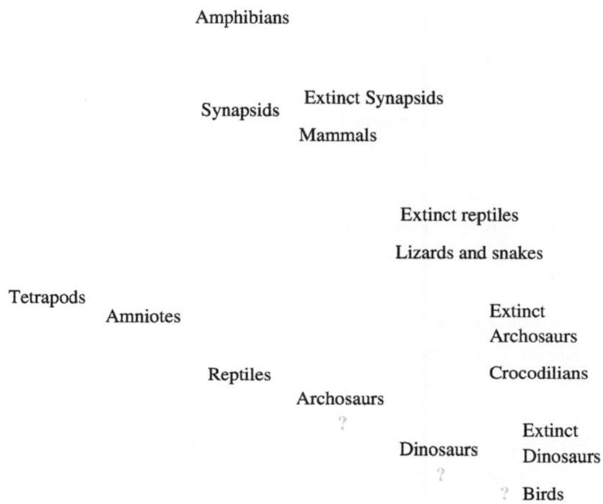

Simple example cladogram

Warm-bloodedness evolved somewhere in the synapsid–mammal transition.

? Warm-bloodedness must also have evolved at one of these points – an example of convergent evolution.[43]

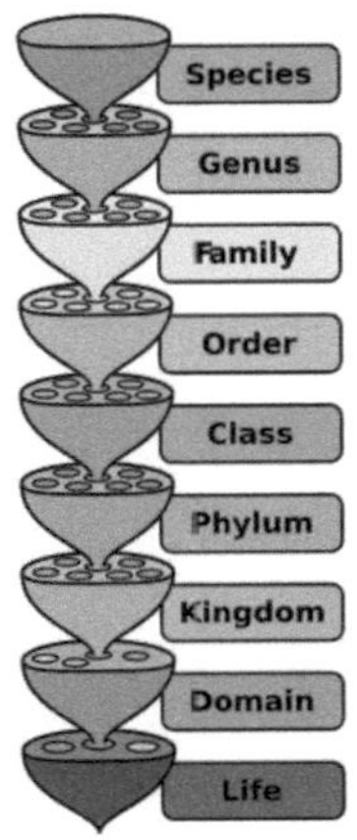

Levels in the Linnean taxonomy

Naming groups of organisms in a way that is clear and widely agreed is important, as some disputes in palaeontology have been based just on misunderstandings over names.[44] Linnean taxonomy is commonly used for classifying living organisms, but runs into difficulties when dealing with newly-discovered organisms that are significantly different from known ones. For example: it is hard to decide at what level to place a new higher-level grouping, e.g. genus or family or order; this is important since the Linnean rules for naming groups are tied to their levels, and hence if a group is moved to a different level it has to be renamed.[45]

Paleontologists generally use approaches based on cladistics, a technique for working out the evolutionary "family tree" of a set of organisms.[44] It works by the logic that, if groups B and C have more similarities to each other than either has to group A, then B and C are more closely related to each other than either is to A. Characters that are compared may be anatomical, such as the presence of a notochord, or molecular, by comparing sequences of DNA or proteins. The result of a successful analysis is a hierarchy of clades – groups that share a common ancestor. Ideally the "family tree" has only two branches leading from each node ("junction"), but sometimes there is too little information to achieve this and paleontologists have to make do with junctions that have several branches. The cladistic technique is sometimes fallible, as some features, such as wings or camera eyes, evolved more than once, convergently – this must be taken into account in analyses.[43]

Evolutionary developmental biology, commonly abbreviated to "Evo Devo", also helps paleontologists to produce "family trees". For example the embryological development of some modern brachiopods suggests that brachiopods may be descendants of the halkieriids, which became extinct in the Cambrian period.[46]

Estimating the dates of organisms

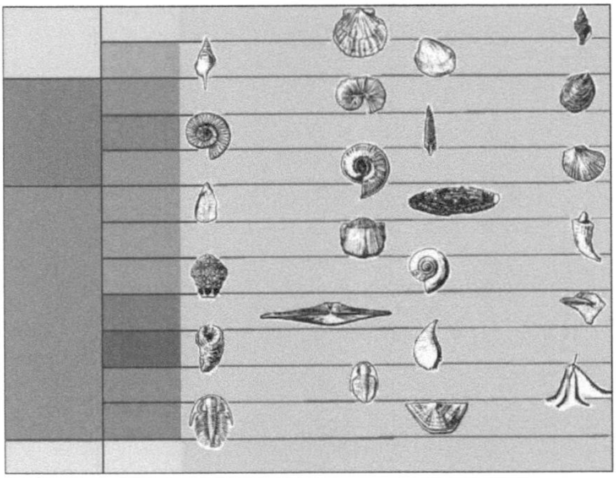

Cenozoic

Mesozoic

Paleozoic

Proterozoic

Quater-
nary

Tertiary

Creta-
ceous

Jurassic

Triassic

Permian

Missis-
sippian

Pennsyl-
vanian

Devo-
nian

Silurian

Ordo-
vician

Camb-
rian

Pecten gibbus

Calyptraphorus velatus

Scaphites hippocrepis

Perisphinctes tiziani

Trophites subbullatus

Leptodus americanus

Cactocrinus multibrachiatus

Dictyoclostus americanus

Mucrospinifer mucronatus

Cystiphyllum niagarense

Bathyurus extans

Paradoxides pinus

Neptunea tabulata

Venericardia planicosta

Inoceramus labiatus

Nerinea trinodosa

Monotis subcircularis

Parafusilina bosei

Lophophyllidium proliferum

Prolecanites gurleyi

Palmatolepus unicornis

Hexamocaras hertzeri

Tetragraptus fructicosus

Billingsella corrugata

Common index fossils used to date rocks in North-East USA

Paleontology seeks to map out how living things have changed through time. A substantial hurdle to this aim is the difficulty of working out how old fossils are. Beds which preserve fossils typically lack the radioactive elements needed for radiometric dating. This technique is our only means of giving rocks greater than about 50 million years old an absolute age, and can be accurate to within 0.5% or better.[47] Although radiometric dating requires very careful laboratory work, its basic principle is simple: the rates at which various radioactive elements decay are known, and so the ratio of the radioactive element to the element into which it decays shows how long ago the radioactive element was incorporated into the rock. Radioactive elements are common only in rocks with a volcanic origin, and so the only fossil-bearing rocks that can be dated radiometrically are a few volcanic ash layers.[47]

Consequently, paleontologists must usually rely on stratigraphy to date fossils. Stratigraphy is the science of deciphering the "layer-cake" that is the sedimentary record, and has been compared to a jigsaw puzzle.[48] Rocks normally form relatively horizontal layers, with each layer younger than the one underneath it. If a fossil is found between two layers whose ages are known, the fossil's age must lie between the two known ages.[49] Because rock sequences are not continuous, but may be broken up by faults or periods of erosion, it is very difficult to match up rock beds that are not directly next to one another. However, fossils of species that survived for a relatively short time can be used to link up isolated rocks: this technique is called *biostratigraphy*. For instance, the conodont *Eoplacognathus pseudoplanus* has a short range in the Middle Ordovician period.[50] If rocks of unknown age are found to have traces of *E. pseudoplanus*, they must have a mid-Ordovician age. Such index fossils must be distinctive, be globally distributed and have a short time range to be useful. However, misleading results are produced if the index fossils turn out to have longer fossil ranges than first thought.[51] Stratigraphy and biostratigraphy can in general provide only relative dating (*A* was before *B*), which is often sufficient for studying evolution. However, this is difficult for some time periods, because of the problems involved in matching up rocks of the same age across different continents.[52]

Family-tree relationships may also help to narrow down the date when lineages first appeared. For instance, if fossils of B or C date to X million years ago and the calculated "family tree" says A was an ancestor of B and C, then A must have evolved more than X million years ago.

It is also possible to estimate how long ago two living clades diverged – i.e. approximately how long ago their last common ancestor must have lived – by assuming that DNA mutations accumulate at a constant rate. These "molecular clocks", however, are fallible, and provide only a very approximate timing: for example, they are not sufficiently precise and reliable for estimating when the groups that feature in the Cambrian explosion first evolved,[53] and estimates produced by different techniques may vary by a factor of two.[21]

Overview of the history of life

The evolutionary history of life stretches back to over 3000 [54] million years ago, possibly as far as 3800 [2] million years ago. Earth formed about 4570 [55] million years ago and, after a collision that formed the Moon about 40 million years later, may have cooled quickly enough to have oceans and an atmosphere about 4440 [56] million years ago.[57] However there is evidence on the Moon of a Late Heavy Bombardment from 4000 to 3800 [58] million years ago. If, as seem likely, such a bombardment struck Earth at the same time, the first atmosphere and oceans may have been stripped away.[59] The oldest clear evidence of life on Earth dates to 3000 [54] million years ago, although there have been reports, often disputed, of fossil bacteria from 3400 [60] million years ago and of geochemical evidence for the presence of life 3800 [2] million years ago.[19] [61] Even the simplest modern organisms are too complex to have emerged directly from non-living materials.[62] Some scientists have proposed that life on Earth was "seeded" from elsewhere,[63] but most research concentrates on various explanations of how life could have arisen independently on Earth.[64]

This wrinkled "elephant skin" texture is a trace fossil of a non-stromatolite microbial mat.
The image shows the location, in the Burgsvik beds of Sweden, where the texture was first identified as evidence of a microbial mat.[65]

For about 2,000 million years microbial mats, multi-layered colonies of different types of bacteria, were the dominant life on Earth.[66] The evolution of oxygenic photosynthesis enabled them to play the major role in the oxygenation of the atmosphere[67] from about 2400 [3] million years ago. This change in the atmosphere increased their effectiveness as nurseries of evolution.[68] While eukaryotes, cells with complex internal structures, may have been present earlier, their evolution speeded up when they acquired the ability to transform oxygen from a poison to a powerful source of energy in their metabolism. This innovation may have come from primitive eukaryotes capturing oxygen-powered bacteria as endosymbionts and transforming them into organelles called mitochondria.[69] The earliest evidence of complex eukaryotes with organelles such as mitochondria, dates from 1850.0 [70] million years ago.[29]

Multicellular life is composed only of eukaryotic cells, and the earliest evidence for it is the Francevillian Group Fossils from 2100 [71] million years ago,[72] although specialization of cells for different functions first appears between 1430 [73] million years ago (a possible fungus) and 1200 [74] million years ago (a probable red alga). Sexual reproduction may be a prerequisite for specialization of cells, as an asexual multicellular organism might be at risk of being taken over by rogue cells that retain the ability to reproduce.[75] [76]

Opabinia made the largest single contribution to modern interest in the Cambrian explosion.

The earliest known animals are cnidarians from about 580 [5] million years ago, but these are so modern-looking that the earliest animals must have appeared before then.[77] Early fossils of animals are rare because they did not develop mineralized hard parts that fossilize easily until about 548 [78] million years ago.[79] The earliest modern-looking bilaterian animals appear in the Early Cambrian, along with several "weird wonders" that bear little obvious resemblance to any modern animals. There is a long-running debate about whether this Cambrian explosion was truly a very rapid period of evolutionary experimentation; alternative views are that modern-looking animals began evolving earlier but fossils of their precursors have not yet been found, or that the "weird wonders" are evolutionary "aunts" and "cousins" of modern groups.[80] Vertebrates remained an obscure group until the first fish with jaws appeared in the Late Ordovician.[81] [82]

The spread of life from water to land required organisms to solve several problems, including protection against drying out and supporting themselves against gravity.[83] [84] The earliest evidence of land plants and land invertebrates date back to about 476 [85] million years ago and 490.0 [6] million years ago respectively.[84] [86] The lineage that produced land vertebrates evolved later but very rapidly between 370 [7] million years ago and 360 [87] million years ago;[88] recent discoveries have overturned earlier ideas about the history and driving forces behind their evolution.[89] Land plants were so successful that they caused an ecological crisis in the Late Devonian, until the evolution and spread of fungi that could digest dead wood.[31]

At about 13 centimetres (5.1 in) the Early Cretaceous *Yanoconodon* was longer than the average mammal of the time.[90]

Birds are the last surviving dinosaurs.[91]

During the Permian period synapsids, including the ancestors of mammals, may have dominated land environments,[92] but the Permian–Triassic extinction event 251.0 [93] million years ago came very close to wiping out complex life.[94] During the slow recovery from this catastrophe a previously obscure group, archosaurs, became the most abundant and diverse terrestrial vertebrates. One archosaur group, the dinosaurs, were the dominant land vertebrates for the rest of the Mesozoic,[95] and birds evolved from one group of dinosaurs.[91] During this time mammals' ancestors survived only as small, mainly nocturnal insectivores, but this apparent set-back may have accelerated the development of mammalian traits such as endothermy and hair.[96] After the Cretaceous–Tertiary extinction event 65 [10] million years ago killed off the non-avian dinosaurs – birds are the only surviving dinosaurs – mammals increased rapidly in size and diversity, and some took to the air and the sea.[97] [98] [99]

A modern social insect collects pollen from a modern flowering plant.

Fossil evidence indicates that flowering plants appeared and rapidly diversified in the Early Cretaceous, between 130 [100] million years ago and 90 [101] million years ago.[102] Their rapid rise to dominance of terrestrial ecosystems is thought to have been propelled by coevolution with pollinating insects.[103] Social insects appeared around the same time and, although they account for only small parts of the insect "family tree", now form over 50% of the total mass of all insects.[104]

Humans evolved from a lineage of upright-walking apes whose earliest fossils date from over 6 [11] million years ago.[105] Although early members of this lineage had chimp-sized brains, about 25% as big as modern humans', there are signs of a steady increase in brain size after about 3 [106] million years ago.[107] There is a long-running debate about whether *modern* humans are descendants of a single small population in Africa, which then migrated all over the world less than 200,000 years ago and replaced previous hominine species, or arose worldwide at the same time as a result of interbreeding.[108]

Mass extinctions

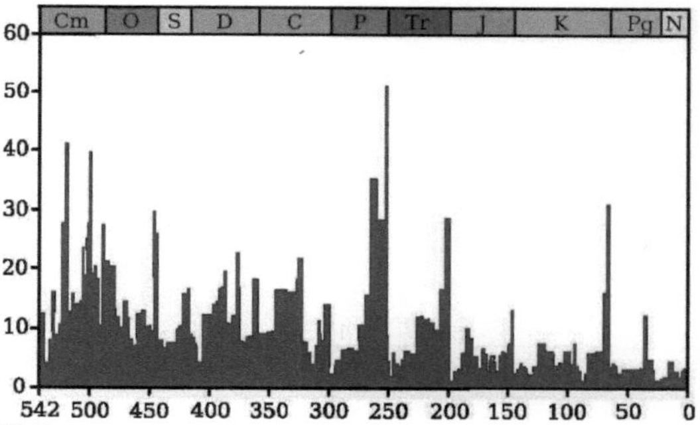

K–T

Tr–J

P–Tr

Late D

O–S

Millions of years ago

Apparent extinction intensity, i.e. the fraction of genera going extinct at any given time, as reconstructed from the fossil record (graph not meant to include recent epoch of Holocene extinction event)

Life on earth has suffered occasional mass extinctions at least since 542 [109] million years ago. Although they are disasters at the time, mass extinctions have sometimes accelerated the evolution of life on earth. When dominance of particular ecological niches passes from one group of organisms to another, it is rarely because the new dominant group is "superior" to the old and usually because an extinction event eliminates the old dominant group and makes way for the new one.[110] [111]

The fossil record appears to show that the rate of extinction is slowing down, with both the gaps between mass extinctions becoming longer and the average and background rates of extinction decreasing. However, it is not certain whether the actual rate of extinction has altered, since both of these observations could be explained in several ways:[112]

- The oceans may have become more hospitable to life over the last 500 million years and less vulnerable to mass extinctions: dissolved oxygen became more widespread and penetrated to greater depths; the development of life on land reduced the run-off of nutrients and hence the risk of eutrophication and anoxic events; marine ecosystems became more diversified so that food chains were less likely to be disrupted.[113] [114]
- Reasonably complete fossils are very rare, most extinct organisms are represented only by partial fossils, and complete fossils are rarest in the oldest rocks. So paleontologists have mistakenly assigned parts of the same organism to different genera which were often defined solely to accommodate these finds – the story of *Anomalocaris* is an example of this.[115] The risk of this mistake is higher for older fossils because these are often unlike parts of any living organism. Many of the "superfluous" genera are represented by fragments which are not found again and the "superfluous" genera appear to become extinct very quickly.[112]

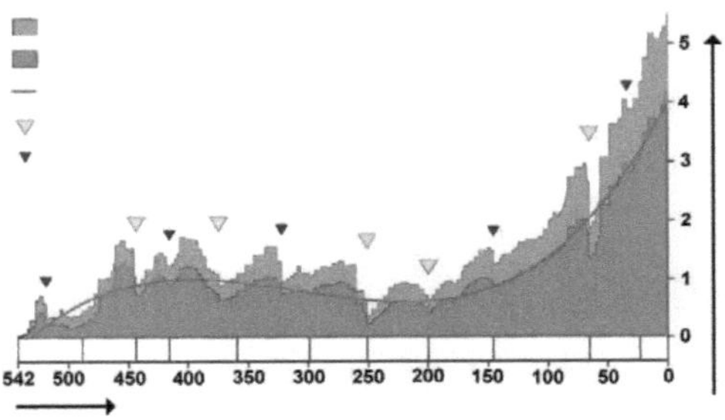

All genera

"Well-defined" genera

Trend line

"Big Five" mass extinctions

Other mass extinctions

Million years ago

Thousands of genera

Phanerozoic biodiversity as shown by the fossil record

Biodiversity in the fossil record, which is

> "the number of distinct genera alive at any given time; that is, those whose first occurrence predates and whose last occurrence postdates that time"[116]

shows a different trend: a fairly swift rise from 542 to 400 [117] million years ago, a slight decline from 400 to 200 [118] million years ago, in which the devastating Permian–Triassic extinction event is an important factor, and a swift rise from 200 [119] million years ago to the present.[116]

History of paleontology

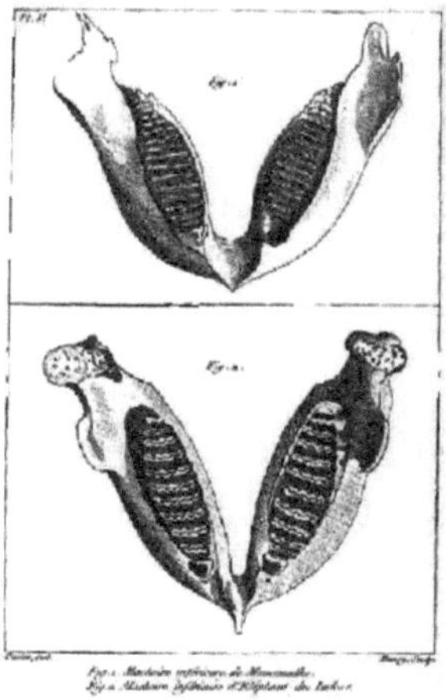

This illustration of an Indian elephant jaw and a mammoth jaw (top) is from Cuvier's 1796 paper on living and fossil elephants.

Although paleontology became established around 1800, earlier thinkers had noticed aspects of the fossil record. The ancient Greek philosopher Xenophanes (570–480 BC) concluded from fossil sea shells that some areas of land were once under water.[120] During the Middle Ages the Persian naturalist Ibn Sina, known as *Avicenna* in Europe, discussed fossils and proposed a theory of petrifying fluids on which Albert of Saxony elaborated in the 14th century.[121] The Chinese naturalist Shen Kuo (1031–1095) proposed a theory of climate change based on the presence of petrified bamboo in regions that in his time were too dry for bamboo.[122]

In early modern Europe, the systematic study of fossils emerged as an integral part of the changes in natural philosophy that occurred during the Age of Reason. At the end of the 18th century Georges Cuvier's work established comparative anatomy as a scientific discipline and, by proving that some fossil animals resembled no living ones, demonstrated that animals could become extinct, leading to the emergence of paleontology.[123] The expanding knowledge of the fossil record also played an increasing role in the development of geology, particularly stratigraphy.[124]

The first half of the 19th century saw geological and paleontological activity become increasingly well organized with the growth of geologic societies and museums[125] [126] and an increasing number of professional geologists and fossil specialists. Interest increased for reasons that were not purely scientific, as geology and paleontology helped industrialists to find and exploit natural resources such as coal.[127]

This contributed to a rapid increase in knowledge about the history of life on Earth and to progress in the definition of the geologic time scale, largely based on fossil evidence. In 1822 Henri Marie Ducrotay de Blanville, editor of *Journal de Phisique*, coined the word "paleontology" to refer to the study of ancient living organisms through fossils.[128] As knowledge of life's history continued to improve, it became increasingly obvious that there had been some kind of successive order to the development of life. This encouraged early evolutionary theories on the transmutation of species.[129] After Charles Darwin published *Origin of Species* in 1859, much of the focus of paleontology shifted to understanding evolutionary paths, including human evolution, and evolutionary theory.[129]

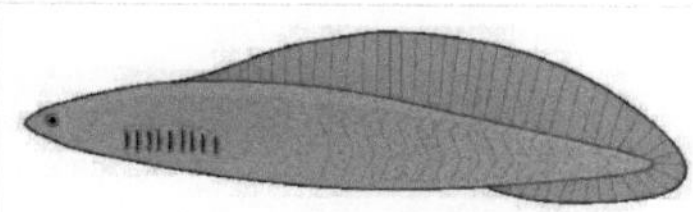

Haikouichthys, from about 518 [130] million years ago in China, may be the earliest known fish.[131]

The last half of the 19th century saw a tremendous expansion in paleontological activity, especially in North America.[132] The trend continued in the 20th century with additional regions of the Earth being opened to systematic fossil collection. Fossils found in China near the end of the 20th century have been particularly important as they have provided new information about the earliest evolution of animals, early fish, dinosaurs and the evolution of birds.[133] The last few decades of the 20th century saw a renewed interest in mass extinctions and their role in the evolution of life on Earth.[134] There was also a renewed interest in the Cambrian explosion that apparently saw the development of the body plans of most animal phyla. The discovery of fossils of the Ediacaran biota and developments in paleobiology extended knowledge about the history of life back far before the Cambrian.[80]

Increasing awareness of Gregor Mendel's pioneering work in genetics led first to the development of population genetics and then in the mid-20th century to the modern evolutionary synthesis, which explains evolution as the outcome of events such as mutations and horizontal gene transfer which provide genetic variation, with genetic drift and natural selection driving changes in this variation over time.[135] Within the next few years the role and operation of DNA in genetic inheritance were discovered, leading to what is now known as the "Central Dogma" of molecular biology.[136] In the 1960s molecular phylogenetics, the investigation of evolutionary "family trees" by techniques derived from biochemistry, began to make an impact, particularly when it was proposed that the human lineage had diverged from apes much more recently than was generally thought at the time.[137] Although this early study compared proteins from apes and humans, most molecular phylogenetics research is now based on comparisons of RNA and DNA.[138]

See also

- Archaeobiology
- Dinosaurs
- Evolutionary history of life
- Fossil collecting
- Fossils
- Geology
- History of paleontology
- Important publications in paleontology
- List of fossil sites *(with link directory)*
- List of notable fossils
- List of transitional fossils
- Radiometric dating
- Taxonomy of commonly fossilised invertebrates
- Timeline of paleontology
- Treatise on Invertebrate Paleontology

Notes

[1] from Greek: *παλαιός (palaeos)* "old, ancient", *ὄν, ὀντ- (on, ont-)* "being, creature", and *λόγος (logos)* "speech, thought"

[2] http://toolserver.org/~verisimilus/Timeline/Timeline.php?Ma=3800

[3] http://toolserver.org/~verisimilus/Timeline/Timeline.php?Ma=2400

[4] http://toolserver.org/~verisimilus/Timeline/Timeline.php?Ma=1700–1200

[5] http://toolserver.org/~verisimilus/Timeline/Timeline.php?Ma=580

[6] http://toolserver.org/~verisimilus/Timeline/Timeline.php?Ma=490

[7] http://toolserver.org/~verisimilus/Timeline/Timeline.php?Ma=370

[8] http://toolserver.org/~verisimilus/Timeline/Timeline.php?Ma=230

[9] http://toolserver.org/~verisimilus/Timeline/Timeline.php?Ma=150

[10] http://toolserver.org/~verisimilus/Timeline/Timeline.php?Ma=65

[11] http://toolserver.org/~verisimilus/Timeline/Timeline.php?Ma=6

[12] Cowen, R. (2000). *History of Life* (3rd ed.). Blackwell Science. p. xi. ISBN 063204444-6.

[13] Laporte, L.F. (October 1988). "What, after All, Is Paleontology?". *PALAIOS* **3** (5): 453. doi:10.2307/3514718. JSTOR 3514718.

[14] Laudan, R. (1992). "What's so Special about the Past?" (http://books.google.com/?id=kyLRtsvLS2AC&pg=PA55&dq="What's+so+Special+about+the+Past"+laudan&q="What's so Special about the Past" laudan). In Nitecki, M.H., and Nitecki, D.V.. *History and Evolution*. SUNY Press. p. 58. ISBN 0791412113. . Retrieved 7 February 2010.

[15] Cleland, C.E. (September 2002). "Methodological and Epistemic Differences between Historical Science and Experimental Science" (http://spot.colorado.edu/~cleland/articles/Cleland.PS.Pdf) (PDF). *Philosophy of Science* **69**: 474–496. doi:10.1086/342453. . Retrieved September 17, 2008.

[16] *McGraw-Hill Encyclopedia of Science & Technology*. McGraw-Hill. 2002. p. 58. ISBN 0079136656.

[17] Laudan, R. (1992). "What's so Special about the Past?". In Nitecki, M.H., and Nitecki, D.V.. *History and Evolution*. SUNY Press. p. 57. ISBN 0791412113.

[18] "How does paleontology differ from anthropology and archaeology?" (http://www.ucmp.berkeley.edu/faq.php#anthro). University of California Museum of Paleontology. . Retrieved September 17, 2008.

[19] Brasier, M., McLoughlin, N., Green, O., and Wacey, D. (June 2006). "A fresh look at the fossil evidence for early Archaean cellular life" (http://physwww.mcmaster.ca/~higgsp/3D03/BrasierArchaeanFossils.pdf) (PDF). *Philosophical Transactions of the Royal Society: Biology* **361** (1470): 887–902. doi:10.1098/rstb.2006.1835. PMC 1578727. PMID 16754605. . Retrieved August 30, 2008.

[20] Twitchett RJ, Looy CV, Morante R, Visscher H, Wignall PB (2001). "Rapid and synchronous collapse of marine and terrestrial ecosystems during the end-Permian biotic crisis". *Geology* **29** (4): 351–354. Bibcode 2001Geo....29..351T. doi:10.1130/0091-7613(2001)029<0351:RASCOM>2.0.CO;2.

[21] Peterson, Kevin J., and Butterfield, N.J. (2005). "Origin of the Eumetazoa: Testing ecological predictions of molecular clocks against the Proterozoic fossil record". *Proceedings of the National Academy of Sciences* **102** (27): 9547–52. Bibcode 2005PNAS..102.9547P. doi:10.1073/pnas.0503660102. PMC 1172262. PMID 15983372.

[22] Hutchinson, John R.; Garcia, M (28 February 2002). "*Tyrannosaurus* was not a fast runner" (http://www.nature.com/nature/journal/v415/n6875/full/4151018a.html). *Nature* **415** (6875): 1018–1021. doi:10.1038/4151018a. PMID 11875567. . Summary in press release No Olympian: Analysis hints *T. rex* ran slowly, if at all (http://www.sciencenews.org/articles/20020302/fob1.asp)

[23] Meers, M.B. (August 2003). "Maximum bite force and prey size of *Tyrannosaurus rex* and their relationships to the inference of feeding behavior" (http://www.ingentaconnect.com/content/tandf/ghbi/2003/00000016/00000001/art00001). *Historical Biology: A Journal of Paleobiology* **16** (1): 1–12. doi:10.1080/0891296021000050755. .

[24] Cady, S.L. (April 1998). "Astrobiology: A New Frontier for 21st Century Paleontologists". *PALAIOS* **13** (2): 95–97. doi:10.2307/3515482. JSTOR 3515482. PMID 11542813.

[25] Plotnick, R.E.. "A Somewhat Fuzzy Snapshot of Employment in Paleontology in the United States" (http://palaeo-electronica.org/2008_1/commentary/comment.htm). *Palaeontologia Electronica* (Coquina Press) **11** (1). ISSN 1094-8074. . Retrieved September 17, 2008.

[26] "What is Paleontology?" (http://www.ucmp.berkeley.edu/FAQ/whatispaleo.html). University of California Museum of Paleontology. . Retrieved September 17, 2008.

[27] Kitchell, J.A. (1985). "Evolutionary Paleocology: Recent Contributions to Evolutionary Theory" (http://www.ucmp.berkeley.edu/FAQ/whatispaleo.html). *Paleobiology* **11** (1): 91–104. . Retrieved September 17, 2008.

[28] Hoehler, T.M., Bebout, B.M., and Des Marais, D.J. (19 July 2001). "The role of microbial mats in the production of reduced gases on the early Earth" (http://www.nature.com/nature/journal/v412/n6844/full/412324a0.html). *Nature* **412** (6844): 324–327. doi:10.1038/35085554. PMID 11460161. . Retrieved July 14, 2008.

[29] Hedges, S.B., Blair, J.E, Venturi, M.L., and Shoe, J.L. (January 2004). "A molecular timescale of eukaryote evolution and the rise of complex multicellular life" (http://www.biomedcentral.com/1471-2148/4/2/abstract/). *BMC Evolutionary Biology* **4** (2): 2. doi:10.1186/1471-2148-4-2. PMC 341452. PMID 15005799. . Retrieved July 14, 2008.

[30] "Paleoclimatology" (http://matrix.geology.ohio-state.edu/ProspectiveGradStudents/school-of-earth-sciences-directory/specialties/global-and-environmental-change/paleoclimatology). Ohio State University. . Retrieved September 17, 2008.

[31] Algeo, T.J., and Scheckler, S.E. (1998). "Terrestrial-marine teleconnections in the Devonian: links between the evolution of land plants, weathering processes, and marine anoxic events". *Philosophical Transactions of the Royal Society: Biology* **353**: 113–130. doi:10.1098/rstb.1998.0195. PMC 1692181.

[32] "Biostratigraphy: William Smith" (http://evolution.berkeley.edu/evosite/history/biostratigraphy.shtml). . Retrieved September 17, 2008.

[33] "Biogeography: Wallace and Wegener (1 of 2)" (http://evolution.berkeley.edu/evosite/history/biogeography.shtml). University of California Museum of Paleontology and University of California at Berkeley. . Retrieved September 17, 2008.

[34] "What is paleontology?" (http://www.ucmp.berkeley.edu/faq.php#paleo). University of California Museum of Paleontology. . Retrieved September 17, 2008.

[35] Benton MJ, Wills MA, Hitchin R (2000). "Quality of the fossil record through time". *Nature* **403** (6769): 534–7. Bibcode 2000Natur.403..534B. doi:10.1038/35000558. PMID 10676959.

Non-technical summary (http://palaeo.gly.bris.ac.uk/cladestrat/news.html)

[36] Butterfield, N.J. (2003). "Exceptional Fossil Preservation and the Cambrian Explosion" (http://icb.oxfordjournals.org/cgi/content/abstract/43/1/166). *Integrative and Comparative Biology* **43** (1): 166–177. doi:10.1093/icb/43.1.166. PMID 21680421. . Retrieved June 28, 2008.

[37] Cowen, R. (2000). *History of Life* (3rd ed.). Blackwell Science. p. 61. ISBN 063204444-6.

[38] Butterfield, N.J. (2001). "Ecology and evolution of Cambrian plankton" (http://66.102.1.104/scholar?hl=en&lr=&ie=UTF-8&q=cache:9xeRu1SdF0QJ:www.earthscape.org/r3/ES14785/ch09.pdf+). *The Ecology of the Cambrian Radiation* (New York: Columbia University Press): 200–216. . Retrieved September 27, 2007.

[39] Signor, P.W. (1982). "Sampling bias, gradual extinction patterns and catastrophes in the fossil record" (http://www.csa.com/partners/viewrecord.php?requester=gs&collection=TRD&recid=A8425675AH). *Geological implications of impacts of large asteroids and comets on the earth* (Boulder, CO: Geological Society of America): 291–296. A 84–25651 10–42. . Retrieved January 1, 2008.

[40] Fedonkin, M.A., Gehling, J.G., Grey, K., Narbonne, G.M., Vickers-Rich, P. (2007). *The Rise of Animals: Evolution and Diversification of the Kingdom Animalia* (http://books.google.com/?id=OFKG6SmPNuUC&pg=PA213&lpg=PA213&dq=trace+fossil+complex+animal). JHU Press. pp. 213–216. ISBN 0801886791. . Retrieved November 14, 2008.

[41] e.g. Seilacher, A. (1994). "How valid is Cruziana Stratigraphy?" (http://www.springerlink.com/index/WP279834395100KH.pdf) (PDF). *International Journal of Earth Sciences* **83** (4): 752–758. . Retrieved September 9, 2007.

[42] Brocks, J.J., Logan, G.A., Buick, R., and Summons, R.E. (1999). "Archaean molecular fossils and the rise of eukaryotes" (http://www.sciencemag.org/cgi/content/abstract/285/5430/1033). *Science* **285** (5430): 1033–1036. doi:10.1126/science.285.5430.1033.

PMID 10446042. . Retrieved September 2, 2008.

[43] Cowen, R. (2000). *History of Life* (3rd ed.). Blackwell Science. pp. 47–50. ISBN 063204444-6.

[44] Brochu, C.A, and Sumrall, C.D. (July 2001). "Phylogenetic Nomenclature and Paleontology". *Journal of Paleontology* **75** (4): 754–757. doi:10.1666/0022-3360(2001)075<0754:PNAP>2.0.CO;2. JSTOR 1306999.

[45] Ereshefsky, M. (2001). *The Poverty of the Linnaean Hierarchy: A Philosophical Study of Biological Taxonomy* (http://books.google.com/?id=tM6E8-_vSD0C&dq="The+Poverty+of+the+Linnaean+Hierarchy:+A+Philosophical+Study+of+Biological+Taxonomy"&pg=PP1). Cambridge University Press. p. 5. ISBN 0521781701. .

[46] Cohen, B. L. and Holmer, L. E. and Luter, C. (2003). "The brachiopod fold: a neglected body plan hypothesis" (http://eprints.gla.ac.uk/2920/01/Cohen_2920.pdf) (PDF). *Palaeontology* **46** (1): 59–65. doi:10.1111/1475-4983.00287. . Retrieved August 7, 2008.

[47] Martin, M.W.; Grazhdankin, D.V.; Bowring, S.A.; Evans, D.A.D.; Fedonkin, M.A.; Kirschvink, J.L. (May 5, 2000). "Age of Neoproterozoic Bilaterian Body and Trace Fossils, White Sea, Russia: Implications for Metazoan Evolution" (abstract). *Science* **288** (5467): 841–5. Bibcode 2000Sci...288..841M. doi:10.1126/science.288.5467.841. PMID 10797002.

[48] Pufahl, P.K., Grimm, K.A., Abed, A.M., and Sadaqah, R.M.Y. (October 2003). "Upper Cretaceous (Campanian) phosphorites in Jordan: implications for the formation of a south Tethyan phosphorite giant". *Sedimentary Geology* **161** (3-4): 175–205. Bibcode 2003SedG..161..175P. doi:10.1016/S0037-0738(03)00070-8.

[49] "Geologic Time: Radiometric Time Scale" (http://pubs.usgs.gov/gip/geotime/radiometric.html). U.S. Geological Survey. . Retrieved September 20, 2008.

[50] Löfgren, A. (2004). "The conodont fauna in the Middle Ordovician *Eoplacognathus pseudoplanus* Zone of Baltoscandia" (http://journals.cambridge.org/production/action/cjoGetFulltext?fulltextid=235488). *Geological Magazine* **141** (04): 505–524. doi:10.1017/S0016756804009227. . Retrieved November 17, 2008.

[51] Gehling, James; Jensen, Sören; Droser, Mary; Myrow, Paul; Narbonne, Guy (March 2001). "Burrowing below the basal Cambrian GSSP, Fortune Head, Newfoundland" (http://www.journals.cambridge.org/action/displayAbstract?fromPage=online&aid=74669). *Geological Magazine* **138** (2): 213–218. doi:10.1017/S001675680100509X. . Retrieved November 17, 2008.

[52] e.g. Gehling, James; Jensen, Sören; Droser, Mary; Myrow, Paul; Narbonne, Guy (March 2001). "Burrowing below the basal Cambrian GSSP, Fortune Head, Newfoundland" (http://www.journals.cambridge.org/action/displayAbstract?fromPage=online&aid=74669). *Geological Magazine* **138** (2): 213–218. doi:10.1017/S001675680100509X. . Retrieved November 17, 2008.

[53] Hug, L.A., and Roger, A.J. (2007). "The Impact of Fossils and Taxon Sampling on Ancient Molecular Dating Analyses". *Molecular Biology and Evolution* **24** (8): 889–1897. doi:10.1093/molbev/msm115. PMID 17556757.

[54] http://toolserver.org/~verisimilus/Timeline/Timeline.php?Ma=3000

[55] http://toolserver.org/~verisimilus/Timeline/Timeline.php?Ma=4570

[56] http://toolserver.org/~verisimilus/Timeline/Timeline.php?Ma=4440

[57] * "Early Earth Likely Had Continents And Was Habitable" (http://www.colorado.edu/news/releases/2005/438.html). 2005-11-17. .
* Cavosie, A. J.; J. W. Valley, S. A., Wilde, and E.I.M.F. (July 15, 2005). "Magmatic $\delta^{18}O$ in 4400-3900 Ma detrital zircons: A record of the alteration and recycling of crust in the Early Archean" (http://www.sciencedirect.com/science?_ob=ArticleURL&_udi=B6V61-4GDKB05-3&_coverDate=07/15/2005&_alid=382434001&_rdoc=1&_fmt=&_orig=search&_qd=1&_cdi=5801&_sort=d&view=c&_acct=C000050221&_version=1&_urlVersion=0&_userid=10&md5=be47e49c535d059be188b66c6e596dd5). *Earth and Planetary Science Letters* **235** (3-4): 663–681. Bibcode 2005E&PSL.235..663C. doi:10.1016/j.epsl.2005.04.028. .

[58] http://toolserver.org/~verisimilus/Timeline/Timeline.php?Ma=4000–3800

[59] Dauphas, N., Robert, F., and Marty, B. (December 2000). "The Late Asteroidal and Cometary Bombardment of Earth as Recorded in Water Deuterium to Protium Ratio". *Icarus* **148** (2): 508–512. Bibcode 2000Icar..148..508D. doi:10.1006/icar.2000.6489.

[60] http://toolserver.org/~verisimilus/Timeline/Timeline.php?Ma=3400

[61] Schopf, J. (2006). "Fossil evidence of Archaean life". *Philos Trans R Soc Lond B Biol Sci* **361** (1470): 869–85. doi:10.1098/rstb.2006.1834. PMC 1578735. PMID 16754604.

[62] Cowen, R. (2000). *History of Life* (3rd ed.). Blackwell Science. p. 6. ISBN 0632044446.

[63] * Arrhenius, S. (1903). "The Propagation of Life in Space". *Die Umschau volume=7.* Reprinted in Goldsmith, D.,, ed. *The Quest for Extraterrestrial Life.* University Science Books. ISBN 0198557043.
* Hoyle, F., and Wickramasinghe, C. (1979). "On the Nature of Interstellar Grains". *Astrophysics and Space Science* **66**: 77–90. Bibcode 1979Ap&SS..66...77H. doi:10.1007/BF00648361.
* Crick, F. H.; Orgel, L. E. (1973). "Directed Panspermia". *Icarus* **19**: 341–348. Bibcode 1973Icar...19..341C. doi:10.1016/0019-1035(73)90110-3.

[64] Peretó, J. (2005). "Controversies on the origin of life" (http://www.im.microbios.org/0801/0801023.pdf) (PDF). *Int. Microbiol.* **8** (1): 23–31. PMID 15906258. . Retrieved October 7, 2007.

[65] Manten, A.A. (1966). "Some problematic shallow-marine structures" (http://igitur-archive.library.uu.nl/geo/2006-1213-210452/UUindex.html). *Marine Geol* **4**: 227–232. doi:10.1016/0025-3227(66)90023-5. . Retrieved June 18, 2007.

[66] Krumbein, W.E., Brehm, U., Gerdes, G., Gorbushina, A.A., Levit, G. and Palinska, K.A. (2003). "Biofilm, Biodictyon, Biomat Microbialites, Oolites, Stromatolites, Geophysiology, Global Mechanism, Parahistology" (http://134.106.242.33/krumbein/htdocs/Archive/397/Krumbein_397.pdf). In Krumbein, W.E., Paterson, D.M., and Zavarzin, G.A.. *Fossil and Recent Biofilms: A Natural History of Life on Earth.* Kluwer Academic. pp. 1–28. ISBN 1402015976. . Retrieved July 9, 2008.

[67] Hoehler, T.M., Bebout, B.M., and Des Marais, D.J. (July 19, 2001). "The role of microbial mats in the production of reduced gases on the early Earth" (http://www.nature.com/nature/journal/v412/n6844/full/412324a0.html). *Nature* **412** (6844): 324–327. doi:10.1038/35085554. PMID 11460161. . Retrieved July 14, 2008.

[68] Nisbet, E.G., and Fowler, C.M.R. (December 7, 1999). "Archaean metabolic evolution of microbial mats" (PDF). *Proceedings of the Royal Society: Biology* **266** (1436): 2375. doi:10.1098/rspb.1999.0934. PMC 1690475.

[69] Gray MW, Burger G, Lang BF (March 1999). "Mitochondrial evolution". *Science* **283** (5407): 1476–81. Bibcode 1999Sci...283.1476G. doi:10.1126/science.283.5407.1476. PMID 10066161.

[70] http://toolserver.org/~verisimilus/Timeline/Timeline.php?Ma=1850

[71] http://toolserver.org/~verisimilus/Timeline/Timeline.php?Ma=2100

[72] El Albani, Abderrazak; Bengtson, Stefan; Canfield, Donald E.; Bekker, Andrey; Macchiarelli, Reberto (July 2010). "Large colonial organisms with coordinated growth in oxygenated environments 2.1 Gyr ago". *Nature* **466** (7302): 100–104. Bibcode 2010Natur.466..100A. doi:10.1038/nature09166. PMID 20596019.

[73] http://toolserver.org/~verisimilus/Timeline/Timeline.php?Ma=1430

[74] http://toolserver.org/~verisimilus/Timeline/Timeline.php?Ma=1200

[75] Butterfield, N.J. (September 2000). "*Bangiomorpha pubescens* n. gen., n. sp.: implications for the evolution of sex, multicellularity, and the Mesoproterozoic/Neoproterozoic radiation of eukaryotes" (http://paleobiol.geoscienceworld.org/cgi/content/abstract/26/3/386). *Paleobiology* **26** (3): 386–404. doi:10.1666/0094-8373(2000)026<0386:BPNGNS>2.0.CO;2. . Retrieved 2008-09-02.

[76] Butterfield, N.J. (2005). "Probable Proterozoic fungi" (http://paleobiol.geoscienceworld.org/cgi/content/abstract/31/1/165). *Paleobiology* **31** (1): 165–182. doi:10.1666/0094-8373(2005)031<0165:PPF>2.0.CO;2. . Retrieved 2008-09-02.

[77] Chen, J.-Y., Oliveri, P., Gao, F., Dornbos, S.Q., Li, C-W., Bottjer, D.J. and Davidson, E.H. (August 2002). "Precambrian Animal Life: Probable Developmental and Adult Cnidarian Forms from Southwest China" (http://www.uwm.edu/~sdornbos/PDF's/Chen et al.2002.pdf) (PDF). *Developmental Biology* **248** (1): 182–196. doi:10.1006/dbio.2002.0714. PMID 12142030. . Retrieved September 3, 2008.

[78] http://toolserver.org/~verisimilus/Timeline/Timeline.php?Ma=548

[79] Bengtson, S. (2004). *Early skeletal fossils* (http://www.cosmonova.org/download/18.4e32c81078a8d9249800021554/Bengtson2004ESF.pdf). In Lipps, J.H., and Waggoner, B.M.. "Neoproterozoic — Cambrian Biological Revolutions". *Paleontological Society Papers* **10**: 67–78. . Retrieved July 18, 2008.

[80] Marshall, C.R. (2006). "Explaining the Cambrian "Explosion" of Animals" (http://arjournals.annualreviews.org/doi/abs/10.1146/annurev.earth.33.031504.103001?journalCode=earth). *Annu. Rev. Earth Planet. Sci.* **34**: 355–384. Bibcode 2006AREPS..34..355M. doi:10.1146/annurev.earth.33.031504.103001. . Retrieved November 6, 2007.

[81] Conway Morris, S. (August 2, 2003). "Once we were worms" (http://cas.bellarmine.edu/tietjen/Evolution/once_we_were_worms.htm). *New Scientist* **179** (2406): 34. . Retrieved September 5, 2008.

[82] Sansom I.J., Smith, M.M., and Smith, M.P. (2001). "The Ordovician radiation of vertebrates". In Ahlberg, P.E.. *Major Events in Early Vertebrate Evolution*. Taylor and Francis. pp. 156–171. ISBN 0-415-23370-4.

[83] Selden, P.A. (2001). ""Terrestrialization of Animals"" (http://books.google.com/?id=AHsrhGOTRM4C&pg=PA71&lpg=PA71&dq="Terrestrialization+of+Animals"+selden). In Briggs, D.E.G., and Crowther, P.R.. *Palaeobiology II: A Synthesis*. Blackwell. pp. 71–74. ISBN 0632051493. . Retrieved September 5, 2008.

[84] Kenrick, P., and Crane, P.R. (September 1997). "The origin and early evolution of plants on land" (http://blog.richmond.edu/paleontology/files/2008/03/nature-evol-plants-kendrickcrane1997.pdf) (PDF). *Nature* **389** (6646): 33. Bibcode 1997Natur.389...33K. doi:10.1038/37918. . Retrieved 2010-11-10.

[85] http://toolserver.org/~verisimilus/Timeline/Timeline.php?Ma=476

[86] MacNaughton, R.B., Cole, J.M., Dalrymple, R.W., Braddy, S.J., Briggs, D.E.G., and Lukie, T.D. (May 2002). "First steps on land: Arthropod trackways in Cambrian-Ordovician eolian sandstone, southeastern Ontario, Canada" (http://geology.geoscienceworld.org/cgi/content/abstract/30/5/391). *Geology* **30** (5): 391–394. Bibcode 2002Geo....30..391M. doi:10.1130/0091-7613(2002)030<0391:FSOLAT>2.0.CO;2. . Retrieved 2008-09-05.

[87] http://toolserver.org/~verisimilus/Timeline/Timeline.php?Ma=360

[88] Gordon, M.S, Graham, J.B., and Wang, T. (September/October 2004). "Revisiting the Vertebrate Invasion of the Land". *Physiological and Biochemical Zoology* **77** (5): 697–699. doi:10.1086/425182.

[89] Clack, J.A. (November 2005). "Getting a Leg Up on Land" (http://www.sciam.com/article.cfm?id=getting-a-leg-up-on-land). *Scientific American*. . Retrieved September 6, 2008.

[90] Luo, Z., Chen, P., Li, G., & Chen, M. (March 2007). "A new eutriconodont mammal and evolutionary development in early mammals". *Nature* **446** (7133): 288–293. Bibcode 2007Natur.446..288L. doi:10.1038/nature05627. PMID 17361176.

[91] Padian, Kevin. (2004). "Basal Avialae". In Weishampel, David B.; Dodson, Peter; & Osmólska, Halszka (eds.). *The Dinosauria* (Second ed.). Berkeley: University of California Press. pp. 210–231. ISBN 0-520-24209-2.

[92] Sidor, C.A., O'Keefe, F.R., Damiani, R., Steyer, J.S., Smith, R.M.H., Larsson, H.C.E., Sereno, P.C., Ide, O, and Maga, A. (April 2005). "Permian tetrapods from the Sahara show climate-controlled endemism in Pangaea" (http://www.nature.com/nature/journal/v434/n7035/full/nature03393.html). *Nature* **434** (7035): 886–889. Bibcode 2005Natur.434..886S. doi:10.1038/nature03393. PMID 15829962. . Retrieved September 8, 2008.

[93] http://toolserver.org/~verisimilus/Timeline/Timeline.php?Ma=251

[94] Benton M.J. (2005). *When Life Nearly Died: The Greatest Mass Extinction of All Time*. Thames & Hudson. ISBN 978-0500285732.

[95] Benton, M.J. (March 1983). "Dinosaur Success in the Triassic: a Noncompetitive Ecological Model" (http://palaeo.gly.bris.ac.uk/Benton/reprints/1983success.pdf). *Quarterly Review of Biology* **58** (1). . Retrieved September 8, 2008.

[96] Ruben, J.A., and Jones, T.D. (2000). "Selective Factors Associated with the Origin of Fur and Feathers" (http://icb.oxfordjournals.org/cgi/content/full/40/4/585). *American Zoologist* **40** (4): 585–596. doi:10.1093/icb/40.4.585. .

[97] Alroy J. (March 1999). "The fossil record of North American mammals: evidence for a Paleocene evolutionary radiation". *Systematic biology* **48** (1): 107–18. doi:10.1080/106351599260472. PMID 12078635.

[98] Simmons, N.B., Seymour,K.L., Habersetzer, J.,and Gunnell, G.F. (February 2008). "Primitive Early Eocene bat from Wyoming and the evolution of flight and echolocation". *Nature* **451** (7180): 818–821. Bibcode 2008Natur.451..818S. doi:10.1038/nature06549. PMID 18270539.

[99] J. G. M. Thewissen, S. I. Madar, and S. T. Hussain (1996). "*Ambulocetus natans*, an Eocene cetacean (Mammalia) from Pakistan". *Courier Forschungsinstitut Senckenberg* **191**: 1–86.

[100] http://toolserver.org/~verisimilus/Timeline/Timeline.php?Ma=130

[101] http://toolserver.org/~verisimilus/Timeline/Timeline.php?Ma=90

[102] Crane, P.R., Friis, E.M., and Pedersen, K.R. (2000). "The Origin and Early Diversification of Angiosperms" (http://books.google.com/?id=ZJe_Dmdbm-QC&pg=PA233&dq=evolution+flowering+plant+angiosperm). In Gee, H.. *Shaking the Tree: Readings from Nature in the History of Life*. University of Chicago Press. pp. 233–250. ISBN 0226284964. . Retrieved September 9, 2008.

[103] Crepet, W.L. (November 2000). "Progress in understanding angiosperm history, success, and relationships: Darwin's abominably "perplexing phenomenon"" (http://www.pnas.org/content/97/24/12939.full.pdf+html). *Proceedings of the National Academy of Sciences* **97** (24): 12939–12941. Bibcode 2000PNAS...9712939C. doi:10.1073/pnas.97.24.12939. PMC 34068. PMID 11087846. . Retrieved September 9 , 2008.

[104] Wilson, E.O., and Hölldobler, B. (September 2005). "Eusociality: Origin and consequences" (http://www.pnas.org/content/102/38/13367.full.pdf–html). *Proceedigs of the National Acedemy of Sciences* **102** (38): 13367–13371. Bibcode 2005PNAS..10213367W. doi:10.1073/pnas.0505858102. PMC 1224642. PMID 16157878. . Retrieved September 7, 2008.

[105] Brunet M., Guy, F., Pilbeam, D., Mackaye, H.T., *et al.* (July 2002). "A new hominid from the Upper Miocene of Chad, Central Africa" (http://www.nature.com/nature/journal/v418/n6894/abs/nature00879.html). *Nature* **418** (6894): 145–151. doi:10.1038/nature00879. PMID 12110880. . Retrieved September 9, 2008.

[106] http://toolserver.org/~verisimilus/Timeline/Timeline.php?Ma=3

[107] De Miguel, C., and M. Henneberg, M. (2001). "Variation in hominid brain size: How much is due to method?" (http://www.sciencedirect.com/science?_ob=ArticleURL&_udi=B7GW4-4DPCHXC-2&_user=10&_rdoc=1&_fmt=&_orig=search&_sort=d&view=c&_version=1&_urlVersion=0&_userid=10&md5=aef79dbca1f189c885cfe9f36636b131). *HOMO — Journal of Comparative Human Biology* **52** (1): 3–58. doi:10.1078/0018-442X-00019. . Retrieved September 9, 2008.

[108] Leakey, Richard (1994). *The Origin of Humankind*. Science Masters Series. New York, NY: Basic Books. pp. 87–89. ISBN 0465053130.

[109] http://toolserver.org/~verisimilus/Timeline/Timeline.php?Ma=542

[110] Benton, M.J. (2004). "6. Reptiles Of The Triassic" (http://www.blackwellpublishing.com/book.asp?ref=0632056371). *Vertebrate Palaeontology*. Blackwell. ISBN 0045660026. . Retrieved November 17, 2008.

[111] Van Valkenburgh, B. (1999). "Major patterns in the history of xarnivorous mammals" (http://arjournals.annualreviews.org/doi/abs/10.1146/annurev.earth.27.1.463). *Annual Review of Earth and Planetary Sciences* **26**: 463–493. Bibcode 1999AREPS..27..463V. doi:10.1146/annurev.earth.27.1.463. .

[112] MacLeod, Norman (2001-01-06). "Extinction!" (http://www.firstscience.com/home/articles/earth/extinction-page-2-1_1258.html). . Retrieved September 11, 2008.

[113] Martin, R.E. (1995). "Cyclic and secular variation in microfossil biomineralization: clues to the biogeochemical evolution of Phanerozoic oceans". *Global and Planetary Change* **11** (1): 1. Bibcode 1995GPC....11....1M. doi:10.1016/0921-8181(94)00011-2.

[114] Martin, R.E. (1996). "Secular increase in nutrient levels through the Phanerozoic: Implications for productivity, biomass, and diversity of the marine biosphere". *PALAIOS* **11** (3): 209–219. doi:10.2307/3515230. JSTOR 3515230.

[115] Gould, S.J. (1990). *Wonderful Life: The Burgess Shale and the Nature of History*. Hutchinson Radius. pp. 194–206. ISBN 0091742714.

[116] Rohde, R.A., and Muller, R.A. (March 2005). "Cycles in fossil diversity" (http://muller.lbl.gov/papers/Rohde-Muller-Nature.pdf) (PDF). *Nature* **434** (7030): 208–210. Bibcode 2005Natur.434..208R. doi:10.1038/nature03339. PMID 15758998. . Retrieved September 22, 2008.

[117] http://toolserver.org/~verisimilus/Timeline/Timeline.php?Ma=542–400

[118] http://toolserver.org/~verisimilus/Timeline/Timeline.php?Ma=400–200

[119] http://toolserver.org/~verisimilus/Timeline/Timeline.php?Ma=200

[120] Rudwick, Martin J.S. (1985). *The Meaning of Fossils* (2nd ed.). The University of Chicago Press. p. 39. ISBN 0-226-73103-0.

[121] Rudwick, Martin J.S. (1985). *The Meaning of Fossils* (2nd ed.). The University of Chicago Press. p. 24. ISBN 0-226-73103-0.

[122] Needham, Joseph (1986). Science and Civilization in China: Volume 3, Mathematics and the Sciences of the Heavens and the Earth. Caves Books Ltd. p. 614. ISBN 0-253-34547-2.

[123] McGowan, Christopher (2001). The Dragon Seekers. Persus Publishing. pp. 3–4. ISBN 0-7382-0282-7.

[124] Palmer, D. (2005). *Earth Time: Exploring the Deep Past from Victorian England to the Grand Canyon*. Wiley. ISBN 780470022214.

[125] Greene, Marjorie; David Depew (2004). *The Philosophy of Biology: An Episodic History*. Cambridge University Press. pp. 128–130. ISBN 0-521-64371-6.

[126] Bowler, Peter J.; Iwan Rhys Morus (2005). *Making Modern Science*. The University of Chicago Press. pp. 168–169. ISBN 0-226-06861-7.
[127] Rudwick, Martin J.S. (1985). *The Meaning of Fossils* (2nd ed.). The University of Chicago Press. pp. 200–201. ISBN 0-226-73103-0.
[128] Rudwick, Martin J.S. (2008). *Worlds Before Adam: The Reconstruction of Geohistory in the Age of Reform*. The University of Chicago Press. p. 48. ISBN 0-226-73128-6.
[129] Buckland W & Gould SJ (1980). *Geology and Mineralogy Considered With Reference to Natural Theology (History of Paleontology)*. Ayer Company Publishing. ISBN 978-0405127069.
[130] http://toolserver.org/~verisimilus/Timeline/Timeline.php?Ma=518
[131] Shu, D-G., Conway Morris, S., Han, J., *et al.* (January 2003). "Head and backbone of the Early Cambrian vertebrate Haikouichthys" (http://www.nature.com/nature/journal/v421/n6922/abs/nature01264.html). *Nature* **421** (6922): 526–529. Bibcode 2003Natur.421..526S. doi:10.1038/nature01264. PMID 12556891. . Retrieved September 21, 2008.
[132] Everhart, Michael J. (2005). Oceans of Kansas: A Natural History of the Western Interior Sea. Indiana University Press. p. 17. ISBN 0-253-34547-2.
[133] Gee, H., ed (2001). *Rise of the Dragon: Readings from Nature on the Chinese Fossil Record* (http://books.google.com/?id=6WYZQwmyWy0C&dq=Rise+of+the+Dragon++By+Henry+Gee&printsec=frontcover). Chicago, Ill. ;London: University of Chicago Press. pp. 276. ISBN 0226284913. . Retrieved September 21, 2008.
[134] Bowler, Peter J. (2003). Evolution:The History of an Idea. University of California Press. pp. 351–352. ISBN 0-52023693-9.
[135] Bowler, Peter J. (2003). Evolution:The History of an Idea. University of California Press. pp. 325–339. ISBN 0-52023693-9.
[136] Crick, F.H.C. (1955). "On degenerate templates and the adaptor hypothesis" (http://genome.wellcome.ac.uk/assets/wtx030893.pdf) (PDF). . Retrieved October 4, 2008.
[137] Sarich V.M., and Wilson A.C. (December 1967). "Immunological time scale for hominid evolution" (http://www.sciencemag.org/cgi/content/abstract/158/3805/1200). *Science* **158** (3805): 1200–1203. Bibcode 1967Sci...158.1200S. doi:10.1126/science.158.3805.1200. PMID 4964406. . Retrieved September 21, 2008.
[138] Page, R.D.M, and Holmes, E.C. (1998). *Molecular Evolution: A Phylogenetic Approach*. Oxford: Blackwell Science. p. 2. ISBN 0865428891.

References

External links

- Smithsonian's Paleobiology website: a good introduction (http://paleobiology.si.edu/)
- University of California Museum of Paleontology FAQ About Paleontology (http://www.ucmp.berkeley.edu/FAQ/faq.html)
- The Paleontological Society (http://www.paleosoc.org)
- The Palaeontological Association (http://www.palass.org)
- The Paleontology Portal (http://www.paleoportal.org)
- International Palaeoentomological Society (http://fossilinsects.net/index.html)

Specific name (zoology)

Specific name in **zoological** nomenclature (also: **specific epithet** or **species epithet**) refers to the second part (the second name) within the name of a species (a binomen). The first part of the name of a species is the name of the genus or the generic name.

The rules and regulations governing the giving of a new species name are explained in the article species description.

> *Example:* The scientific name for humans is *Homo sapiens*, which is the species name, consisting of two names: *Homo* is the "generic name" (the name of the genus) and *sapiens* is the "specific name".

The grammar of species names

Grammatically, a binomen (and a trinomen, also) must be treated as if it were a Latin phrase, no matter which language the words were originally taken from. (This gives some justification to the popular usage of the phrase "Latin name" instead of the more correct phrase "scientific name".) Grammatically (using Latin grammar), the specific name must adhere to certain conventions. The specific name can either be formed as:

- **A noun in apposition to the genus name**, for example, the scientific name of the lion, *Panthera leo*. In these cases, the word for the genus and the word for the species do not necessarily have to agree in gender. Species names which are nouns in apposition are sometimes the vernacular name of the organism in Latin or Ancient Greek, or the name (specific or generic) of another organism which the organism itself resembles.
- **A noun in the genitive case** (i.e. belonging to).
  - This is common in parasites: *Xenos vesparum* ("Xenos of the wasps").
  - Proper nouns which are names of people and places are often used in the genitive case. For example the name of the coelacanth, *Latimeria chalumnae* which means "Latimeria of Chalumna", is a reference to the area near the mouth of the Chalumna River in the Indian Ocean, where the coelacanth was first found, i.e. its type locality.
- **An adjective** which must agree in case and gender with the genus *Felis silvestris* ("the forest cat")

Differences from botany

In botanical nomenclature, "name" *always* refers to the *whole name* (of a species or otherwise), whereas in zoological nomenclature it can refer to either part of the binomen. Thus *Hedera helix* (Common Ivy, English Ivy) is the name of the species; *Hedera* is the name of the genus; but *helix* is the specific epithet, *not* the specific name.[1]

References

[1] McNeill, J.; Barrie, F.R.; Burdet, H.M. et al., eds. (2006), *International code of botanical nomenclature (Vienna Code) adopted by the seventeenth International Botanical Congress, Vienna, Austria, July 2005* (http://ibot.sav.sk/icbn/main.htm) (electronic ed.), Vienna: International Association for Plant Taxonomy, , retrieved 2011-02-20, Article 23.1 (http://ibot.sav.sk/icbn/no frames/0027Ch3Sec4a023.htm)

- Banks, Nathan (2009). *The Entomological Code. a Code of Nomenclature for Use in Entomology*. BiblioBazaar. ISBN 1113266775.
- Matthews, Janice R.; Bowen, John M.; Matthews, Robert W. (2000). *Successful scientific writing: a step-by-step guide for the biological and medical sciences*. Cambridge University Press. p. 176. ISBN 0521789621.

Article Sources and Contributors

Draconyx *Source*: http://en.wikipedia.org/w/index.php?title=Draconyx *Contributors*: Cave User, Chunky Rice, DeadEyeArrow, Delldot, Firsfron, J. Spencer, Jerkov, Johan Moberg, Kumioko, MWAK, Madchester, Pedroalexandrade, Rnnsh, Smeira, ZayZayEM, 8 anonymous edits

Genus *Source*: http://en.wikipedia.org/w/index.php?title=Genus *Contributors*: 16@r, 5 albert square, A Macedonian, Abanima, Addionne, Aitias, Ajraddatz, Alansohn, Alex.muller, Alexei Kouprianov, AlexiusHoratius, Andre Engels, AndrewWTaylor, AnonMoos, Anonymous Dissident, Anthony Appleyard, Aranea Mortem, Arichnad, Art LaPella, AstroNomer, Athaler, AugPi, Austroungarika, Autonova, AxelBoldt, Baranxtu, Belfry, Benjamnjoel2, Berria, Berserkerz Crit, Berton, Betterusername, Bjankuloski06en, Bobblewik, Bobo192, Bongwarrior, Brian0918, Brianga, Brya, Bryan Derksen, Byron755, CWY2190, Cburnett, Christiangrann, Circeus, CrazyUkee, Cxz111, DMacks, Damnreds, Darth Panda, Daviddariusbijan, Dendrid, DerHexer, Don2k11, Dr CyCoe, Drew roy, Drmies, Dysmorodrepanis, ENeville, ERcheck, ESkog, El C, Eloquence, Emma1700, Enigmaman, Eog1916, Epbr123, Erebus Morgaine, Escape Orbit, Ethien, Firsfron, Flewis, Flyset, Frankenpuppy, Fred Bauder, FreplySpang, FrozenPurpleCube, GB fan, GKoUtSo201, Galoubet, Gdr, GerardM, Giftlite, Gilliam, Glenn, GoEThe, GrahamBould, Graminophile, Grendelkhan, Grunny, Gwern, Hallows AG, Hamamelis, Harloshaply, Hayabusa future, Hbent, Heracles31, Hozro, Hyacinth, I got fived!, Ian Dalziel, Icairns, Ignitus, Iliketochangearticles, Innotata, Invertzoo, Iridescent, J.delanoy, JHolman, Jackfork, Jackjackjr, Japanese Searobin, JemGage, Jerzy, Joblax, John KB, Joseph Solis in Australia, Jossi, Jrockley, Julesd, Julian Mendez, JustAGal, Katimawan2005, Kazenteshi, Kchishol1970, Keilana, KeithTyler, Kesuari, Kimse, Kintaro, Kironide, Kjkolb, Kmsiever, Knutux, Komododragonfan16, Kubigula, Kukini, Kungfuadam, LOL, Lankiveil, Lantianer, Lerdsuwa, Lightmouse, Luckas Blade, MER-C, ML5, MPF, MWAK, Madhero88, Maelin, Maildej, Mani1, Marek69, Maristoddard, Marshman, Mauler90, Mav, Merovingian, Meteor2017, Mfwitten, Mgiganteus1, MichaelSeemann, Mikael Häggström, Mike Dillon, Mike Searson, Mikepenny01, MisfitToys, Mjster, Mmcknight4, Ms2ger, Muhaha, Mukkakukaku, Nakon, NewEnglandYankee, Nihiltres, Nivlak, Nnof002, Nsaa, Nutzmonkey, Oneiros, Orphan Wiki, Pageleft, Paul August, Paul-L, Pengo, Philip Trueman, PierreAbbat, Pinethicket, Pippu d'Angelo, Prashant.saxena, R.e.b., Ram-Man, RazorICE, Redaktor, RexNL, Rich Farmbrough, Richard New Forest, Richard0715, Rjwilmsi, Robert Foley, RobertG, Romanskolduns, Ronhjones, Rossco89, RoundAbout949, Rvollmert, S3000, SP-KP, Salamurai, Sarahcwilkins, Satellite9876, Scientizzle, Sciurinæ, Seahorseruler, Shirt58, Shulei-shulei, SimonP, Sluzzelin, Smith609, Snek01, Snowmanradio, Spiff, Srios653, Standproud1, Starks, Stemonitis, Super Mario, TUF-KAT, Taztouzi1, The Earwig, The Final Chronicler, Tide rolls, Tobias Bergemann, Tommy2010, TotoBaggins, Unara, Uncle G, Unyoyega, Uscurnow17, UtherSRG, Velho, Verbum Veritas, Victor falk, Vietbio, Vina, Vishnava, Vsmith, We.eter, Werts335, Whosyourjudas, Wilbozz, Williamb, Wisco, WolfmanSF, YellowMonkey, Yerpo, Yonwe, Yosri, YuKornilev, Zumbo, Саша Стефановић, 349 anonymous edits

Dinosaur *Source*: http://en.wikipedia.org/w/index.php?title=Dinosaur *Contributors*: *Kat*, *drew, 1-is-blue, 10metreh, 10outof10die, 20075896cu, 2038432048234, 216.175.251.xxx, 2468whodowe, 28421u2232nfenfcenc, 3dnatureguy, 4444hhhh, 5dollabill, 83d40m, 96.149, A Macedonian, AMC0712, Aadal, Aaron Schulz, Aaron Walden, Abigail-II, Abyssal, Academic Challenger, Aceofelves, Adam78, Additional110, Adrian J. Hunter, Againme, Ahoerstemeier, Ajp2000, Albertonykus, Ale jrb, Aleron235, Alex Bakharev, AlexKepler, Aliter, Alpha3103, AmberUTS, Amberrock, Amicon, Amillar, AmyRowe, AnakngAraw, Ancheta Wis, Andre Engels, AndreNatas, Andrew Levine, AndrewHarvey4, Andrewlp1991, Andrewpmk, Andries, Andy, Andycjp, Angado keitho, Anger22, Anonymous editor, Antandrus, Anthony Appleyard, Aquishix, Aragupta, Archangel x105, Archfiendweazal, ArdcoreSoxx, ArielGold, Arpingstone, Art LaPella, ArthurWeasley, Artisticmerit, Aryeztur, Ashcraft, AssistantX, Astro$01, AstroHurricane001, Astropithicus, Athf1234, Attilios, Atulsnischal, Atun, Aunt Entropy, Autodidactyl, Ayrton Prost, Azhyd, B&W Anime Fan, B1atv, BAFAphenom7, BD2412, BWCNY, Babylittletree, Ballista, Bambadil, Banuthev, BartekChom, Basilicofresco, Bassem18, Bbatsell, Bcollecutt, Begoon, Ben Hateva, Ben Jos, Ben m, Benne, Bensaccount, Bento00, Bergsten, Best Known, Bevo, BhaiSaab, Bhumiya, Bidabadi, Bignole, Bill the Duck, Black Kite, Blackicehorizon, Bloodredrover, Blue520, Bob the Wikipedian, Bobber1, Bobblewik, Bobo192, Bodnotbod, Bomac, Bonesiii, BorgHunter, Brad101, BrendanRyan, Brendanconway, Brhoops1, Brian0918, Brighterorange, Brnstormersfan, Bryan Derksen, Bubalana, Buchanan-Hermit, Buckshot128, Buickid, Bulltastrophy, Butros, Buwayahman, Bwilliams, Bwmodular, C4h5n3o, CBDunkerson, CJLL Wright, Caltas, Camerong, Can't sleep, clown will eat me, Canadian-Bacon, CanadianCaesar, Canderson7, Captain Mar vell, Captain02, Captainbeefart, Casliber, Cateybug, Cavrdg, Cbdorsett, Cbogart2, Cdc, Cedrus-Libani, Celestianpower, Cenarium, Centrx, Cephal-odd, Chanting Fox, Chaos, Chowbok, Chris 73, Chris Capoccia, ChrisGriswold, Christian List, Christopherlin, Chuck Marean, Cimp3, Circeus, Cleverclogs06, ClockworkSoul, ClockworkTroll, Cmanoli, Colin stuart, Colonies Chris, CommonsDelinker, Comp25, ConfuciusOrnis, Conkyworm, Connormah, Conti, Conversion script, Cosmo1, Cpandtle, Crag, Craigsjones, Crot buer, Crotalus horridus, Csaw, Curps, Cutter, Cvene64, Cwruidth, Cyberevil, CyclePat, Cyclopia, D, D-Notice, DARTH SIDIOUS 2, DJ Clayworth, DLJessup, DMacks, DSedrez, DVD R W, DabMachine, Dan D. Ric, Danarmak, Daniel J. Leivick, Daniel.Cardenas, DanielCD, Danny, DarkMasterBob, Darwin's Waiting Line, Dave souza, David Fuchs, David Latapie, David Little, Dbachmann, Debivort, Deckiller, Decltype, Degina2008, DeinonychusDinosaur999, Dendodge, Denovoid, Derek Yoda's friend, Deror avi, Desmay, Devotchka, Diberri, DietCaffeineFreeCherryCoke, Digitalme, Dinoguy2, Dinohunter, Dinosauria-Freak, Dinosaurs12345, Discordance, Discospinster, Diucón, Diyako, Dj Capricorn, Dlohcierekim's sock, Doc Tropics, DoctorLarry, Dolovis, Dom Kaos, Domminico, Dr.Bastedo, Dracontes, Dragon Helm, DragonflySixtyseven, Draicone, Drakesiphon, DreamOfMirrors, Dromaeosaur Dude, DropDeadGorgias, Dropzink, Drunkenmonkey, Dudo2, Dysepsion, Dysprosia, ESkog, EWS23, Ed Cormany, Eeeasterbunny, ElTyrant, ElZorroLivesForever, Eleassar, Elembis, Elkester, Elliskev, Elmosdapimp, Eloquence, Eluchil404, Elvarg, Encephalon, Ensrifraff, Epbr123, Eras-mus, Espi, Estel, Evercat, Everyking, Everything212, Evice, EvilZak, Ezra Wax, FF2010, Fama Clamosa, Fan-1967, Farsight001, Fastfission, Fastzander, Fat kidddd, Favyan, Fconaway, Felix Dance, Fenevad, FiGGy-06, Finalius, Firsfron, Flamingspinach, Florentino floro, Fotoguzzi, Freakofnurture, Frecklefoot, Fredrik, Frencheigh, Frosty0814snowman, Fruge, FunkMonk, Funnybunny, Futuremyst, Fuzheado, Fyunck(click), GDallimore, GGordonWorleyIII, GHe, GSGSGSG, GSlicer, Gabbe, Gaga654, Gaia5074Q, Gaius Cornelius, Galassi, Garion96, GarrettD, Garry Denke, Gavin the Chosen, Gazza69, Gazzster, Gene Nygaard, Geologyguy, Gfoley4, Gfuzz, Giftlite, Gilgamesh, Gilliam, Gimlifangirl, Gimmetrow, Givenarmy, Glenn, Globalmuseum, Gloomy Storm, Gnomehead, Gnostrat, Goggle archives, Gogo Dodo, GrAveTzT, GraemeL, Grafen, Graham87, GrahamN, Graminophile, Greanak, GreatWhiteNortherner, Greebo cat, GregRustFan, Gregalodon, Grundle2600, Grutter, Gsklee, Gtrmp, Guambomb, Guybrush, Gwernol, H0merg0mez, H2g2bob, HKMP5A2, HMallison, Hadal, Haeleth, Hall Monitor, Hamiltondaniel, Hans Dunkelberg, Happyguy123, Happyguy32, Happysailor, Harley peters, Harold137, Harrisonjason, Hastyo, Heial123456, Henryisamazingyeh, Henzel, Hermione1980, Herp Derp, Heteren, HiDrNick, Hibernian, Hiddensonyvaio, Hires an editor, HistoryBA, Hmacmil2, Hmains, Hobartimus, HonouraryMix, HoodedMan, Hordaland, Hoseumou, Hottentot, HunterX, Hut 8.5, I 0wn in rs, I teh yuh, IVAN3MAN, Ialsoagree, Icalanise, Ikh, Improv, Indosauros, Infraredeclipse, Innotata, Inventedgravity, InverseHypercube, Invincible Ninja, Iota, Irbnspgn, Irixman, Irvin calicut, Islandbaygardener, Isomorphic, Its me, all feakin me, Itsmine, J. Spencer, JCarriker, JForget, JK43, JOSHKIN, JRLivesey, JW1805, Jacek Kendysz, Jacob.jose, Jafeluv, Jameswilson, JamieSwan1993, Jamotive, Jannex, Jaranda, JarlaxleArtemis, Jason Potter, Jaxl, JayC, Jayen466, Jc-S0CO, Jchap, Jckrull, Jedi of redwall, JeffStickney, JeffTL, Jefffire, Jens Lallensack, JeremyA, Jerkov, JerryOrr, Jfg284, Jiang, Jidanni, Jiddisch, Jim62sch, Jimp, JinJian, Jipcy, Jiy, Jkbjlfk, Jklamm, Jklin, Jmv2009, Jmw0000, Jni, JoeSmack, Joeyrockyhorror, Johann Wolfgang, John Doe or Jane Doe, John Reid, John of Reading, John.Conway, JohnWayne, Jokaer, Jokkel, Jonas Poole, Jorjepot, Josh Grosse, JoshuaZ, Jossi, Joyous!, Jpcoles, Jpo, Jrsnbarn, Jtkiefer, JudahH, Julia W, Jumacdon, Jumbo Snails, Jusdafax, JustinWick, Jwissick, Jyril, Jóna Þórunn, K, Kaijucole, Kandar, Kaoruchan, Karam.Anthony.K, Karimarie, Kazooieman, Kazvorpal, Kbdank71, Kchishol1970, Kdakin, Kent Wang, Kevin B12, Kevindavid, Kglavin, Khoikhoi, Kickaha, Kilgore Sprout, Killdevil, KillerChihuahua, Kilo-Lima, Kintaro, Kipala, Kirill Lokshin, Kirkseeray, Kit Foxtrot, Kleopatra, Klonimus, Kmccoy, KnightRider, Knowledge Examiner, Knowledge Seeker, KnowledgeOfSelf, Knutux, Koliri, Kotengu, Kozuch, KrAtul, KrakatoaKatie, KritonK, Kronermark, Kubigula, Kuru, Kurykh, Kwamikagami, Kwiki, Kyle Barbour, Kyle sb, L Kensington, La goutte de pluie, Lacrimosus, Lanasa, LandonChastelhassy, Landroo, Latka, Lavateraguy, Lbaum, LeeWilson, Leeteditmastorz, LeonWhite, LeonardoRob0t, Leptictidium, Lepton6, Leroyinc, Lexi Marie, Lightmouse, Ligulem, Like omg its karma, LilHelpa, Liridon, Lizhia, Llull, Lord Seth, LordAmeth, Lot49a, Lowellian, Ludwigs2, Lugnuts, Luna Santin, Lupin, Luuva, M C Y 1008, MCBOB, MER-C, MPF, MSGJ, MWAK, MZMcBride, Maastrictian, Mac Dreamstate, MacTire02, Mackinaw, Madchester, Madtast, Magister Mathematicae, Mahanga, Majungatholus, MakeChooChooGoNow, MalK`k, Malo, Mandyjb, Manufracture, MarkBuckles, MarkGallagher, MarkSweep, Markeer, Markjreed, MarvintheParanoidAndroid, Massimo Macconi, Master of Puppets, MasterOfHisOwnDomain, Masterv88, Matt Crypto, Matt McIrvin, Matt me, Matt1314, MattDP, MattKingston, Mattbrown04, MatthewAJYD, Mausy5043, Mav, Max Schwarz, Maximus Rex, Maxthefax123, Mazza024, Mckay history, Mdd4696, Mebden, MechBrowman, Medeis, Mediatenz, Meggar, Melaen, Melesse, Menchi, Mephistophelian, Meridas, Mermaid from the Baltic Sea, Mgglaser, Mgiganteus1, Mgumn, Michael Devore, Michael Zimmermann, Michael-david Reisman, Michaelmas1957, Michaelritchie200, Michaelrogerson, Mike Dillon, Mike Rosoft, MikeLynch, MilkshakeMan246, Millahnna, Minna Sora no Shita, Misza13, Mjbaldwin, Mkdw, Mn extra, Mo0, Mokele, Mokwella, Mollwollfumble, Monicasdude, Moon&Nature, Moreschi, Mormegil, Morrad, Moshe Constantine Hassan Al-Silverburg, Movementarian, Movieresearch, Mr w, Mrdallaway, Mrmanager07, Mschel, Mschlindwein, MuDavid, Murtasa, My76Strat, Myfretsareonfire, Mysid, Mzmadmike, Nakon, Name Theft Victim, Narayanese, Natgoo, NatureA16, NawlinWiki, Nbarth, Neale Monks, Nemokid1220, Netsnipe, Neutrality, Nick Number, Nick UA, Nick125, Niczar, Nikodemos, NinjaCharlie, Ninjatacoshell, No substitute for you, Noah Lawrence, Nodosaurus, Noirish, Not an anon anymore, Nrg800, NuclearWarfare, Nuno Tavares, Nyy2135525, Oblivious, Obradovic Goran, Ocaasi, Ohnoitsjamie, Okapi, Old Father Time, OldakQuill, Oldshonen, Ollivier, Omicronpersei8, Orangemarlin, OrgasGirl, Orie1061, Otolemur crassicaudatus, Ours18, Ovons, Owen, PDH, PJS102, Palica, Pamri, Panzuriel, Paranoid, Parsa, Patrickncollin1, Patricksmullen, Paul Erik, Paul Magnussen, Paul Stansifer, Pcb21, Peachsncream, Pegship, PenguinJockey, Penguinpablo, Perstar, PeruvianIlama, Peter Isotalo, Petri Krohn, Petter Bøckman, Pfhreak, Pgk, Phaedriel, Phatom87, Phifer784, Phil Boswell, Phil1988, Philcha, Philg88, Phlebas, Phoenix79, PiCo, PiMaster3, PiRSquared17, Pickelbarrel, Picone, Pie4all88, Piloscia, Pilotguy, Pinethicket, Pjrich, Pleebloo, Poison the Well, Pomeapplepome, Porqin, Portillo, Predator, PrestonH, Pretty Green, Primate, Pufnstuf, Quadell, Quantum bird, Quercusrobur, R. Baley, R.D.H. (Ghost In The Machine), RG2, RPlunk2853, RWFanMS, Rahimbakhsh, Ramdrake, Rampart, Randomblue, Raptorfact, Raul654, Raven in Orbit, Rdsmith4, Reach Out to the Truth, Rebecca Bangay, RedSpruce, RedWolf, Reddish, Redthoreau, Reedy, RetiredUser2, Retran, Revansrangers, Revolución, RexNL, Riana, Rich Farmbrough, Richard Keatinge, Ringsjöodjuren, Rintrah, Ritchy, Rje, Rjwilmsi, Rm w a vu, Roaribeadino, Rob117, RobertG, RoboAction, Rocastelo, Rogor, Roke, Roleplayer, Rossnixon, Rothorpe, Roy da Vinci, RoyBoy, Royalguard11, Rsm99833, Rufous-crowned Sparrow, Russell C. Sibley, RyanGerbil10, Ryanaxp, Ryanfrm, Ryanmcg2006, Ryoung122, Ryulong, Ryuuseipro, SFH, Sai2020, SallyForth123, Sam Hocevar, Sampi, Samsara, Sango123, Sardanaphalus, Saucerottia, Scary dragon atop the hill, Sceptre, SchuminWeb, Scm83x, Screwwithu99, Scroner, Scruffian, Seaphoto, Seb az86556, Sengkang, Serendipodous, ShadowMan1od, Shadowjams, Shagie, Shaki997, Shanel, Shanes, Shantavira, Sheeana, Sheep81, Sheridan, Shizhao, Shoaler, Siblin, Silverwood, SimonP, SineWave, Siriusblack902, Sjakkalle, Sjorford, Skizzik, SlimVirgin, Smith609, Sneaky Oviraptor18, Snowolf, SoCalSuperEagle, Soap, Someone56, Spacemanjohn, Spangineer, Sparky the Seventh Chaos, SparrowsWing, Spawn Man, Spinodontosaurus, Spinosaur, Spinxy, SpookyMulder, Sports are my world, Spotty11222, Springnuts, Sputnikcccp, Srleffler, Srnec, Sstty4, Starexplorer, Stbalbach, Steinbach, Steinsky, Stemonitis, Stephan Schulz, Stephen G. Brown, Stephen Gilbert, Stephen e nelson, Stephenb, Steven Walling, Steveoc 86, Stickman sam, StoptheDatabaseState, Streona, Stubblyhead, Studio34, Stuff8, Subidei, Sucatraps, SunnyDisp, SuperHamster, Superraptor, Supreme Deliciousness, SusanLarson, Swatjester, Systolic, T.lancer, Tab439, Tannin, Tapir Terrific, TashTish, Tbhotch, Tdlugolecki, TeaDrinker, Tekana, Temtem, TestPilot, Thanatosimii, The Librarian At Terminus, The Man in Question, The Rambling Man, The Ronin, The Singing Badger, The Thing That Should Not Be, The Utahraptor, TheBigJabroni, TheCryingofLot49, TheDevilsTrill, TheFixer000, TheJC, TheKMan, TheProject, TheTruthiness, Theda, ThinkBlue, This user has left wikipedia, Thumperward, Thunderhead, Tide rolls,

TigerShark, TimBentley, TimVickers, Timotheus Canens, Tiptoety, Titanium Dragon, Titoxd, Tjs003, Tjunier, Tktktk, Tmangray, Tmoney792, Tmopkisn, Tobby72, Toffelginkgo, Tom-, TomStar81, Tomalak geretkal, Tommulliner, Tommy2010, Tomsega, Tone, Tony Fox, TonyClarke, Tosoft, Tothebarricades.tk, Tpbradbury, Train2104, Traveler100, Travmitz, TreasuryTag, Trekphiler, Trevor Andersen, Trevor MacInnis, Triceratops, Trojanhero, Troodon58, Tropylium, Trovatore, Truthteller, Trödel, Tsunamicarlos, Tsunomaru, Tuckwai, Tuohirulla, Tyrannosaurus roll, Tyrol5, Unclericо89, UndeadMonkey, Undress 006, UtherSRG, VAcharon, Vague Rant, Vanished User 0001, Vanished user, Vanished user 03, VanishedUser314159, Vargklo, VegaDark, Veledan, Vendetta21, Vera.tetrix, Vessss, Vikreykja, Vina, Violetriga, Vsmith, W33nie, WAS 4.250, WaffleMafia, Wafulz, Waggers, Waker913, Warfwar3, Wars, WarthogDemon, Wavelength, Wayward, Wbroun, WereSpielChequers, Weregerbil, Wesley J M, Westvoja, Wetman, Whatiguana, Whatthree16, Whimemsz, Whooper, Wiggles69, Wik, Wiki alf, Wikipager, Wikipediarules2221, Wikivandalizer, Wildnox, WillNess, William Avery, Wilson44691, Wimt, Winston365, Wkey, WolfmanSF, Woohookitty, Worldtraveller, Wowaconia, Xaosflux, Xerxes314, Yamamoto Ichiro, Yasharelliin, Yes234, Yggdrasilsroot, Yooden, Yossarian, Yummifruitbat, Zaffo, Zaharous, Zanimum, ZeWrestler, Zfervent, Zharta, ZimZalaBim, Zoologyteacher, Zoso95, Zsinj, Zywxn, Ævar Arnfjörð Bjarmason, 1067 anonymous edits

Jurassic *Source*: http://en.wikipedia.org/w/index.php?title=Jurassic *Contributors*: .marc., 10dxholcomb, 16@r, 1exec1, 28421u2232nfenfcenc, 88995ting, A Karley, AMK152, Abce2, Abyssal, Adamsachs, Adrian, Agathman, Ahoerstemeier, Aitias, Alansohn, AlexiusHoratius, Amaltheus1, Amcbride, Ammonoid, Andre Engels, Andrea105, Andrewhermes, Ano-User, Anonymous editor, Antandrus, Anthony Appleyard, Apollonius 1236, Arakunem, Arkuat, Art LaPella, Ascánder, Aughannah, Awickert, Ballista, Banaticus, Bejnar, Benosaurus, Betterusername, Biglatrell, Bill37212, BinaryTed, Blood sliver, Bluefist, Bob232, Bobly3, Bobo192, Bongwarrior, Bookandcoffee, Booksworm, Boom boom guy, Brianchasejared, Bringing, Bryan Derksen, Bucephalus, Burntsauce, Cam, Capricorn42, Captain-tucker, Carcharoth, Casito, Catapult, Ched Davis, Chermundy, ChrisSk8, Chrishowardowns, Chrno Raptor, CiaPan, Civil Engineer III, Cleared as filed, Connormah, Conversion script, Courcelles, Cubanfreak94, Cureden, Cyan, DJ Clayworth, DanielCD, Davidprior, Deflagro, DerHexer, Discospinster, Divercol, Dlloyd, DocWatson42, Don Kenney, Don4of4, Download, Downtown dan seattle, Dr Schrodinger, DrFO.Jr.Tn, Dragon Helm, Drakesiphon, Dreadstar, Drlittletwerp, DuncanHill, Dusik, Dutzi, Dysepsion, Eddy Martin5, El C, El Cid, Elassint, Emperorbma, Enzo Aquarius, Epbr123, Eric B. and Rakim, Erimus, Eyu100, Falcon8765, Fama Clamosa, Fan-1967, Fang 23, Finalnight, Firsfron, Fiskehaps, FloreatAntiquaDomus, Flowerparty, Franco3450, GHe, GLaDOS, Genjix, Geologyguy, Gjd001, Glenn, Gliese876, Gogo Dodo, Goodparley, Graham87, GregMinton, Hadal, Hamtechperson, Hockeyman291, Holder, Hroðulf, IRP, Ian Pitchford, Imc, Immunize, Iridescent, It Is Me Here, IvanLanin, Ixfd64, J, J. Spencer, J.delanoy, Jackvon, Janet1983, Jc-S0CO, Jimp, Jinko48, Jjron, Joao Xavier, JodyB, Josh Grosse, Joy, Jsonitsac, Juliancolton, Jyril, Ka Faraq Gatri, Kev.M.T, Kilagria, KnowledgeRequire, Kristen Eriksen, Kumioko, L Kensington, Lear's Fool, LeaveSleaves, Leolaursen, Leptictidium, Lerdsuwa, Literacola, M Alan Kazlev, MJ94, MKoltnow, Marauder40, Mav, Maximillion Pegasus, Mhese, Michael C Price, Michaelg98, Mikenorton, Miss Madeline, Modulatum, Moon&Nature, MrBoo, Muriel Gottrop, NHRHS2010, Nagy, Natl1, NatureA16, Nivix, Nsaa, Nuge, Nuno Tavares, Nurg, Obersachse, Omicronpersei8, OobadoobaRoBeRt, Orangemarlin, Orion11M87, Oxymoron83, P.Geol, Paranomia, PatGallacher, Pegship, Peko2, Pfranson, Phe, Phil Bridger, Philip Trueman, Physchim62, Pieguydude, Pigslookfunny, Pilotguy, Pinethicket, Pinkunicorn, Polarbear97, Possum, Prolog, Pvasiliadis, Qfl247, Quercusrobur, RadicalOne, RedWolf, Rich Farmbrough, Richtom80, Rjwilmsi, Roadahead, Romanm, Ronhjones, Ryulong, SEWilco, SamNeill TreCool, Sammysocks, Savant13, Secret Squirrel, Siim, Slightsmile, SlimVirgin, Slumjimamp, Slowking Man, Smith609, SnappingTurtle, Some jerk on the Internet, Sophie means wisdom, Spongefrog, Spotty11222, SquidSK, Stephen C. Carlson, Stephenb, Suisui, Sweet xx, THEN WHO WAS PHONE?, Tabletop, Tannin, Temporarily Insane, TheNewPhobia, TheOtherJesse, Thingg, Think outside the box, Tide rolls, Tiptoety, Tohd8BohaithuGh1, Tommy2010, Tree Biting Conspiracy, Tuxlie, Tyler, Udonknome, Ulioceras, UnitedStatesian, Unyoyega, Urosp, UtherSRG, Vanished 6551232, Veesicle, Vegetationlife, Vina, Voyagerfan5761, Vsmith, Vuong Ngan Ha, Wetman, Wiki alf, WildWildBil, Willking1979, Wilson44691, Winchelsea, Wknight94, Wodawik, WolfmanSF, Woohookitty, Wooster09, Wwoods, Xerocs, Yahussain, Yochingy, Zachbelcher, Zfr, ZooFari, 719 anonymous edits

Ornithopod *Source*: http://en.wikipedia.org/w/index.php?title=Ornithopod *Contributors*: Abyssal, Albertonykus, Bernie bernbaum, Bobblehead, CFLeon, Dinoguy2, Dudo, Extremophile, Fabien1309, Fakirbakir, Firsfron, Frencheigh, FreplySpang, FunkMonk, GCarty, Hoseumou, J. Spencer, Jackfork, Jeffmedkeff, Jimp, John.Conway, Kevmin, KnightRider, Kumioko, Kwamikagami, MWAK, Mgiganteus1, MisterSheik, Nicke L, Okedem, Palica, Phlebas, Quintote, Rnnsh, Scarian, Sfrandzi, Square87, Zvar, قيناريد دابع, 23 anonymous edits

Miguel Telles Antunes *Source*: http://en.wikipedia.org/w/index.php?title=Miguel_Telles_Antunes *Contributors*: Bgwhite, Bobtheseventeenth13, Burmeister, Epbr123, Felix Folio Secundus, Neptune5000, Open2universe, Pedroalexandrade, Tabletop, Waacstats, 4 anonymous edits

Octávio Mateus *Source*: http://en.wikipedia.org/w/index.php?title=Oct%C3%A1vio_Mateus *Contributors*: Aricomte, CaughtLBW, Epbr123, Lenin and McCarthy, Magioladitis, Omateus, Paleo71, Pedroalexandrade, Rjwilmsi, Sevela.p, Shatteredeuphoria, 5 anonymous edits

Lourinhã Municipality *Source*: http://en.wikipedia.org/w/index.php?title=Lourinh%C3%A3_Municipality *Contributors*: Firsfron, Fsouza, GiovanniS, Husond, Jetman, Joaopais, Koshelyev, Kwamikagami, Mário, NawlinWiki, Nricardo, Omateus, Pedroalexandrade, Plasticspork, Ron B. Thomson, TigreTiger, Topbanana, Ulric1313, 18 anonymous edits

Portugal *Source*: http://en.wikipedia.org/w/index.php?title=Portugal *Contributors*: *Paul*, -- April, .lll., 05wykesa, 10metreh, 16@r, 23prootie, 24fan24, 28421u2232nfenfcenc, 334a, A More Perfect Onion, Aarktica, Aaron Schulz, Abdullahkj, Abercrombieboy229, Acalamari, Acdar, Acerperi, Acoeano, Acroterion, Adam10101, Adartists, Adashiel, Addshore, AdjustShift, Aec is away, Aezram, Agander2, Ahoerstemeier, Ahuskay, Aitias, Ajaxmaster, Ajdmfvss, Ajrsferreira, Akamad, Akanemoto, Akapulkooo, AlJoseph, Alai, Alansohn, Alarob, Albertc1990, Albertocchaves, Aleenf1, Alex earlier account, Alex.tan, Alex45, Alexander Domanda, Alexbatko, AlexiusHoratius, Alfio, AlfredG, Algebra, Alison, AlistairMcMillan, Allstarecho, Alpha x, Alrise, Alucard (Dr.), Alvaro pacheco, Alvesgaspar, Amafreak123, Amakuru, Amazonien, Amestre, Amfincher, Anabus, Ancheta Wis, Andhanq, Andre Engels, AndreNatas, AndrewHowse, AndrewWTaylor, Andrewpmk, Andrewrp, Andris, Andy Marchbanks, Andy120, Andycor, AngelOfSadness, Angela, Angeliumpt, Angelus boy, Anger22, Angusmclellan, Animalia555, Antandrus, Anthropologique, Apresubu, Apulana, Aranherunar, Archfalhwyl, Arcimpulse, Arcot, Arctic Night, Aris Katsaris, Arjun01, Armendariz, Armindo, ArmsHeldOut, Armydepot, Arnon Chaffin, Arnoutf, Asidemes, Aston martin-jacko, Astral, Atlantis Hawk, Atoric, AuburnPilot, Aude, Augustes, Auntof6, Avala, Avenged Eightfold, Avicennasis, Avril23, Awaterbuffalo, AxG, Ayls, AzaToth, B.Lameira, BBird, BILLYBLAZES247, BLueFiSH.as, BRG, Babel2675, Backpack kiwi, Bahnfrend, BalkanFever, Ballestrain, Bambuway, Banaticus, Barneca, Baronnet, Barryob, Barticus88, Barvinok, Bastin, Batimantel, BattyMan105, Bazonka, Bcnviajero, Bcorr, Beagle398, Beantown 1995, Bear Stearned, Beetstra, Belard, Ben Ben, Benc, Benson85, BernardaAlba, Bettymnz4, Bhadani, Bibliomaniac15, Big Smooth, Bigno1998, Binksternet, Bjarki S, Bkell, Bkwillwm, BlaiseFEgan, Blake-, Bleh999, Blinkms, Blizkrieg, Bloodshedder, Blue-Haired Lawyer, BlueMars, Bluedenim, Bob269, BobMaestro, Bobak, Bobblewik, Bobo192, Boing! said Zebedee, Bonadea, Bongwarrior, BorgQueen, Boris Allen, Bornhj, Bort08, Bowlhover, Brainboy109, Branddobbe, Brazillianmaster, Brighterorange, Brim, Brion VIBBER, Brockert, Bruce1ee, Brutaldeluxe, Bryan Derksen, BryanMar1, Bsapp7, Btyner, Buaidh, Bubba hotep, Burnwelk, BurtAlert, Butterscotch, Byubryguy, C777, CJ, CL, Cacophony, CactusWriter, Cailil, Caknuck, Calafalas, Calhi415, CalicoCatLover, Califate123, Califate123!, Call me Bubba, Calmer Waters, Caltas, Calvin 1998, Camarinha, CambridgeBayWeather, Can't sleep, clown will eat me, CanadianLinuxUser, Cantus, Carbonix, CardinalDan, Cargofast03, Carlwest, Carreira, Casg, Catgut, Ccaravello, Cctoide, Cdani, Cdc, Ceartas, Celtmist, Cenarium, Centrx, Cfailde, Cgid, ChKa, Chairman S., Chanheigeorge, Chanting Fox, CharlotteWebb, Chase me ladies, I'm the Cavalry, Chavash, Checco24, Chembody, Chensiyuan, Chi-editrix, Chicken Wing, Chochopk, Cholmes75, Chris Roy, Chun-hian, Chzz, Ciaccona, Ciceronl, Cigsandalcohol, Ckatz, Cleduc, Closedmouth, Cmdrjameson, Cogito ergo sumo, Colchicum, Colibri37, Colonel beauregard, CommonsDelinker, Conorobradaigh, Conscious, Conversion script, Cool3, Coolhawks88, CopperMurdoch, Corlyon, Corvokarasu, Costa1143, Couto 01, Cptnono, Cracker017, CristianChirita, Crystalclearchanges, Cs-wolves, Curps, Cutdwnthetrees, CyanMoon, Cybercobra, D, DJ Bungi, DMacks, DSRH, DSuser, DUBJAY04, DVdm, Damnreds, DanKeshet, DanMS, Dane 1981, Danger, Danny, Danny Rathjens, DannyMackeycoolkid, Dark Shikari, DarkAudit, Darrendeng, DarthShrine, Dasani, Dasdi, Dataranking, DaveBurstein, DaveJ7, Daven200520, Davewild, David Be, David Gerard, David Kernow, David Liuzzo, Davidcannon, DavisBurke, Dawn22, Dbfirs, Dbiel, Dbtfz, Dcoetzee, Dddstone, Deadlock, Deepblue9000, Deflective, Deirda, Delduca3, Delldot, Dentren, Deon Steyn, Deor, DerHexer, Derek Ross, Desiphral, Devyn339, Dfrg.msc, Diana306, Diannaa, Diogo serrao, Diogo sfreitas, DireWolf, Discospinster, Distal24, DivineIntervention, Dlohcierekim, Doco, Docu, Dodgeviper8000, Doktor Mephisto, Dolive21, Domino theory, Donald Albury, Dontbeakakke, Doric Loon, Dosai, Dougofborg, Doulos Christos, Download, Dr Gangrene, Dr Sachs, DrBob, Dracontes, Drbug, Drew R. Smith, Dumbo1, Duunlaen, Dweller, Dysprosia, E Pluribus Anthony, E lee, E-boy, E2eamon, East37, Eboracum, Eclecticology, EdBever, Edgar2894, Edivorce, Edolen1, EdwinCasadoBaez, Eggstasy, Egil, Egregio, EivindJ, El C, El Ogre, El0i, Elapsed, Electionworld, Elf, Elizabethhaydee, Ellbeecee, Ellmist, Elmondo21st, ElockidAlternate, Emaeka123, Emmalicous(emmy), Empire4, EmpireKing, EncycloPetey, Encyclopedist, Enzino, Epbr123, Epf, Eprojectatuni, Erauch, Erfurt150, Eric Shalov, EricSRodrigues154, Escape Orbit, Esnaz, Etams, EugeneZelenko, Evercat, Everyking, EvocativeIntrigue, Ewlyahoocom, Ewulp, Excelsior Deo, Excirial, Explendido Rocha, Eyadhamid, Eyezanatar, FF2010, FMSousa, Fabrictramp, Falcon8765, Falcorian, Fami619, Faradayplank, Favonian, Feitclub, Fender 01, Feydey, Ffpfp, Ficcionista, Fieldday-sunday, Fifo, Figureground, Filipe24, FilipeS, Fireaxe888, Fishonastick, Fl, Flamarande, Flaquito76, Flatterworld, Flauto Dolce, Fluffernutter, Flyguy649, Flyrs, Fornost, Fortes, Fpenteado, FrancisTyers, FrancoGG, Francolam303, FrankCostanza, Franmen, Fratrep, Freakofnurture, Fredrik, Funeral, Fys, Fæ, G.-M. Cupertino, G51, GTBacchus, Gabbe, Gabhala, Gadfium, Gail, Gaius Cornelius, Galati, Gangb0nga, Ganymead, Gawaxay, Gaz, Gazech90, Gazilion, Gentil79, Geoff Plourde, Georgios, Getmoreatp, Gggh, Ghakko, Ghepeu, Ghost bombs, Ghostbubbles, Gibmetal77, Gidonb, Gilliam, Gimmetrow, Ginajanota, Gioto, Givenchy de Paris, Giving is good, as long as you're getting, Glahglahglah, Glen, Glenn, Gmaxwell, Gnevin, Goblin, Goeagles4321, Gogo Dodo, Golbez, Goldman07, Gonvelho, Good Olfactory, Googooblahblah, GordyB, Gottago, Goustien, Gozar, Gozza gogi, Gr8opinionater, Gracenotes, Graeme Bartlett, GraemeL, Graham87, Gramaic, Grandma945, Greatestrowerever, Green Giant, Greenbreen, Gregathec, Gregoriol619, Grendelkhan, Grim Reaper, Ground Zero, Gryffindor, Gscshoyru, Gtg204y, Guaca, Guidinha, Guilherme Paula, Guinnesscub, Gurch, Gurchzilla, Gz33, Gzdavidwong, Gznorneplatz, Gzornenplatz, HWSager, Haakon, Hakyle, Halfsnail, Hamza best3, HappyJ-man, Harry491, HarryHenryGebel, Harryboyles, Hauser, Hawkestone, Hayden120, Hdt83, Headbomb, HeartofaDog, HeckXX, Hector.C.Jorge, Hehuhihoha, Heimstern, HelderM, Helixdq, Helldisplay, Hennessey, Patrick, HexaChord, HighKing, Highvale, Hillary Duffles, Historian19, History Packet, Hmains, Hofbauer, Holy Santa, Hottentot, Howcheng, Hrohej5y984093f, Hu12, Hughk-h, Hugoxyz, Hurricane111, Husond, Hut 8.5, Hybrid-2k, Hydao, II MusLiM HyBRiD II, Iamaboyyes, Iamthemurmur, Iamtheonewhois, IanManka, Ianb, Ice Cold Beer, Icedust7, Icestorm815, Idontknow610, Ief, Igor nav, Ikheetcalvin, Ikiroid, Ilmari Karonen, ImperatorExercitus, Improv, Indoles, Iner22, Inferno, Lord of Penguins, Infrogmation, Interiot, Investneur, Iol, Iridescent, Isewell, Islaosh, Italiano111, Itsmine, Ivirivi00, Ixfd64, Izaakb, Izzedine, J.delanoy, J04n, JCarriker, JDamanWP, JForget, JIMTHEBURGER, JJGD, JNW, JPEsperanca&FernandaCorreia, JStephanieApril, JW1805, JaGa, Jacek Kendysz, Jackol, Jahnavi7, Jakkaru Kitsune, JamesAM, JamesBWatson, Jamesontai, Jamigi, Jammygit09, Jan Van Leer, Jan1nad, Janneman, Jaranda, Jasonfitz, JavierMC, JayLeno175, Jayden54, Jaygirl193, Jcaraballo, Jeff G., Jeff Silvers, Jeff3000, Jek555, Jeltz, Jensboot, JeremyA, JeremyTraub, JesseGarrett, Jesusalentejo, Jfucinos, Jguk, Jhendin, Jiang, Jiddisch, Jimfbleak, Jimp, Jj137, Jjacobsmeyer, Jjhcap99, Jkliff, Jklin, Jmwe, Jnestorius, JoanneB, Joao, Joaoalve, Joaocastro, Joaodiascoelho, Joaopais, Joeporeflag, JoeyCr7, Joffeloff, John, John of Reading, JohnCD, Johnbibby, Johnsoolaimon, Jomifica, Jomig, Jomilindo, JonHarder, Jonathan.s.kt, Jonik, Joobnob, Jopacs, Jordanmueller, Jorge A. e Baltazar, Jorunn, Jose Manuel, Jose77, Joseolgon, Joseph Solis in Australia, JosephDosSantos, Joshbuddy, Joshua, JoshuaD1991, Jossi, Joy, Joyous!, Jpgordon, Jrbobrj, Jserranog, Jtalledo, Jtkiefer, Jusjih, Justlettersandnumbers, Jvmills, Jweiss11, Jyavner, Jza84, Kaisershatner, Kaldari, Kandaurov, Kangalheiro, Kanguru007, KapilTagore, Karam.Anthony.K, Karl, Katefan0, Keilana, KeithB, Kelly Martin, Kendrick7, Kennethmac2000, Kenshin, Kerotan, Kesac, Kevin B12, Kevin Saff, Kevinkor2, Kgorman-ucb, Khoikhoi, Khukri,

Kimiko, Kimmykun, KingOfSports22, Kingfish, Kingofham, Kingpin13, Kintetsubuffalo, KirrVlad, Kirube, Kitya.ferrer, Kman543210, Kneiphof, Knobbishly, Knutux, Koenige, Kotniski, Koyaanis Qatsi, KoyaanisQatsi, Kozuch, Kreb, Kross, Krun, Ksu1101, Kukini, Kungfuadam, Kuru, Kusma, Kuzwa, Kwamikagami, Kwekubo, Kww, La Pianista, Lacobrigo, Lady Tenar, Lagalag, Lahiru k, Langston, Lankiveil, Lapis, Largoplazo, Latka, Laughing Man, Lbdearmas, Lcarscad, Lear 21, LeaveSleaves, Legolas1094, LeoO3, Leonardo Alves, Levineps, Lfdneves, Lhimec, Liambobjoe, Licor, Liene, Liftarn, Lightmouse, Lights, Ligulem, LilHelpa, Lilac Soul, Lilwikitits, Lindmere, Liquidcentre, Livajo, Lkarno, Lm123lm, Lms, Lockesdonkey, Lolaandmel, London Hawk, LonelyMarble, Lord Dandy, Lord KRISHNA, LordStanley, Losdiablossonaqui, Loukinho, Luis rib, Luis wiki, Luis.frps, Luisvareladasilva, Luna Santin, Lunettes, Lupin, Lusci, Lynntoniolondon, M-Silva91, M. Bernardo, M4corridor1, MECU, MER-C, MIP, MJCdetroit, MMSequeira, MONGO, MZMcBride, Mac Davis, Macanckor, Macarion, Macl, Macy, Madjazzman9134, Mafmafmaf, Magioladitis, Mailer diablo, Majalinno, Makedonas, Malafaya, Malhonen, Malik09, Malo, Man vyi, Mandarax, Mangojuice, Manji23, Manuel Menezes de Sequeira, Manuel de Sousa, Maoraz2001, Marcavis, Marco Neves, Marek69, Marijuanarchy, Mark, Mark Ryan, MarkSutton, MarkSweep, Markalex, Martinwilke1980, Masdawwg, Master Jay, Master of Puppets, Mastercheif, Materialscientist, Matthew Fennell, Matthewrbowker, Mattisse, Mav, Maxgisfat, Maximus Rex, Maxis ftw, Maycoll F. Vieira, Mboverload, McDogm, Mck321, Mcornelius, Mddake, Meaghan, Meandlook, Mesgul82, Metagraph, Metron, MetsFan76, Meursault2004, Mfa fariz, Mgiganteus1, Mhking, Mic, Michael.risley, Michaluis, Mickey gfss2007, Mifa718, Miguel in Portugal, Miguels, Miguelzinho, Mike Rosoft, Milesflint, MindstormsKid, MinnesotanConfederacy, Miranfree, Mjmcb1, Mm40, Mobile Snail, Modernist, Modulatum, Montrealais, MontseBL, Monty845, MoogleEXE, Moralis, Mormegil, Moroccan Spaniard, Moroyou, Morwen, Mouramoor, Mr Stephen, Mr. Joca, Mr45acp, MrBlonde, Mrfish33, Mrshaba, Mschiffler, Mtayl2, Mtord, Muriel Gottrop, Muriga, Murraybuckley, Muéro, Mxn, MyaTya, Mário, N5iln, NCDane, NHRHS2010, Nabla, Nailer123, Nakon, Namiba, Naohiro19, Narrasawa, Naryathegreat, Navy Blue, NawlinWiki, Ndteegarden, Necris, Neilbeach, NeonMerlin, NetBMC, Neutrality, Neverquick, NewEnglandYankee, Nichalp, Nick, Nick cooper UK, Nickin, Nicksss93, Nicksukh, Nickyfann, NigelR, Nightstallion, Nilfanion, Njaelkies Lea, Nmpenguin, Nn123645, No Guru, Nohoguy, Norm, NotAnonymous0, NotJackhorkheimer, Notheruser, Nouse4aname, Novice7, Nricardo, Nsaa, NuclearVacuum, NuclearWarfare, Nuno Gabriel Cabral, Nuno Tavares, Nuno.sanches, Nuttycoconut, Nyttend, O Fenian, ObtuseAngle, Obviamente, Octane, Oda Mari, Ohconfucius, Ohnoitsjamie, Okta87, Oldelpaso, Olessi, Oli.morris@hotmail.co.uk, Olivier, OllieFury, Omicronpersei8, Ooeettuu, Opelio, Opinoso, Optakeover, Optimist on the run, Ordep, Orphan Wiki, OsvaldoGago, Ourique1139, Owen, Owen and eric, Oxymoron83, PFHLai, PZFUN, PaD, PaePae, Page Up, Paine Ellsworth, Panwan, Parent5446, ParkerHiggins, Part Deux, Pastafarian Nights, Patka, Paul Benjamin Austin, Paulluckman, Paulo Oliveira, Paulo oia, Paxsimius, Pc13, Pcdc, Pds0101, Pedro, Pedro Aguiar, PedroPVZ, PedroPalhoto, Pedrodelarosa, Pedrojoper, Pedroshin, PeepP, Peeperman, Peguim, Pekinensis, Penfish, Pensionero, Penwhale, Peregrine981, Pero51621, Persian Poet Gal, Pessim444, Pete de beer@yahoo.com, Peter Horn, PeterStJohn, PeterWild, Pethr, Pethrus, Petzl, Pevarnj, Pgan002, Pgk, Pharaoh of the Wizards, PhilKnight, Philip Trueman, Philip200291, PhillipSaav, Phlegat, Phydend, Physchim62, Picachu, Picapau amarelo, Picapica, Piccolo Modificatore Laborioso, Pigman, Pilim, Pilotguy, Pinethicket, Pingveno, Pinkman45, Piotrl6, Pizza2damaXX, Pmlineditor, Poccil, Poetaris, Polamalu306, Polaron, Poli, Polon11, Poopoonanny, Popdelart, Porkoribs, Portugueseempire, Ppntori, Pro bug catcher, Profession, PrologFan, Proofreader77, Prvc, Psv@1, Puchiko, Pularoid, Pupster21, Pursey, Pwentzel, Quadell, Queson, Quibik, Quintote, Quirky, Quizimodo, Quizimodo (usurped), RA0808, RAMA, RAPCracked, RJII, RJN, RJaguar3, RRParry, Racolaco, Radon210, Rafa28 alq, Ragusan, Rama's Arrow, Ramirez72, Ran, Rande M Sefowt, Rarelibra, RaulCovita, Raven in Orbit, Raven21niner, Ray Cherry, Rbarreira, Rdsmith4, Realinto, Reaper Eternal, Rebel, Recognizance, Red Winged Duck, RedHillian, RedWolf, Redking7, Redlentil, Rednblu, Redraiders203, Reedy, Rei-artur, Reinthal, RekishiEJ, Renato 1998, Rettetast, RexNL, ReyBrujo, Rgoodermote, Rhamiel, Rhialto, Riana, Ricardo-Quaresma, Rich Farmbrough, Rick Block, RickK, Ricklaman, Rinkidinki, Rita.dos.santos, Rjwilmsi, Robert25, Roke, Romanm, Ronark, Ronhjones, Ronline, Rossen4, Rossj81, Rotoblues, Rotring, Rror, RubencinMdE, Ruddyell, Rui.marinheiro, Ruivilao, Ryulong, SJP, SNIyer12, SOAD KoRn, ST47, SU Linguist, SURIV, Saforrest, Saftorangen, Sagasaga, Sailor-Drum, Sam Hocevar, Sam Li, Sam90yo, SamEV, Samantha555, Samrolken, Samuel, Samwb123, Sandover, Sandstein, Sandwich Eater, Sango123, Sanmartin, Sannse, Saoshyant, Sarefo, Scanos, Scgtrp, Schaengel89, Scheridon, Schmendict, SchnitzelMannGreek, Scimitar, Scipius, Sciurinæ, ScottW, ScottishGunner, Sdrtirs, Sean O'Driscoll, Seano1, Secfan, Secretlondon, Seivad, Semperf, Seniorfox, Sennsa, SeoMac, Serte, Señor Juan, Shadowjams, Shanes, Sharkface217, ShelfSkewed, Shibboleth, Shinhan, Shinmawa, Shinpah1, Shirulashem, Shoeofdeath, Shoessss, Showersap12, Shroudan, Sietse Snel, Signsolid, Sillyfolkboy, SimonD, SimonP, SimonX, Sintaku, Sir Lord Emperor, Sir Nicholas de Mimsy-Porpington, Sjakkalle, Skcpublic, Skilgannon, Skizzo3, Skurinski, Slaggart, Slakr, Sleptrip, Smadge1, Smalljim, Smiddle, Smilesfozwood, Smomo, Snigbrook, SnowFire, Snowmanradio, Snowolf, SoSaysChappy, Soap, Sobretudo, Softy, Soman, Some jerk on the Internet, SomeHuman, Sonoflusus, Sonyack, Sophistry, Sortior, SpNeo, SpaceFlight89, SparrowsWing, Spartan-James, Spazzymike1994, SpookyMulder, SpuriousQ, SqueakBox, Stab-in-the-dark, Startstop123, Steel, Stemonitis, Stephenb, Storm Rider, Straatmeester, Struway2, Studerby, Styath, Subspace1250, Subtropical-man, Supertask, Supposed, SusanaM, Svinde, SwisterTwister, Syimrvm, SyntaxError55, T B Pereira, TAG-A-b10, TShilo12, TUF-KAT, Tabletop, Tacv, Tango, Targeman, Tatarian, Taxiguy92, Taxman, Tbhotch, Tcncv, Tcoelho, TedE, Teddycom, Tedzsee, Teixant, Template namespace initialisation script, TenOfAllTrades, Terfili, TerriersFan, ThaddeusB, The Ogre, The Person Who Is Strange, The Rambling Man, The Random Editor, The Red Hat of Pat Ferrick, The Thing That Should Not Be, The Utahraptor, The god lollipop4, The russian leader, TheBlueKnight, TheDoober, TheFelican, TheGeekHead, TheIntersect, TheKMan, ThePointblank, TheTrojanHought, Thehelpfulone, Therequiembellishere, Thewikipedian, Thingg, ThisIsMyUsername, Thisisborin9, ThomasK, Thorius Maximus, Thricecube, ThwartedEfforts, TiagoPereira, Tiagoplb, Tibullus, Tide rolls, TigerShark, Tigerassault, TimComm, Timir2, Timshiels, Tinctorius, TintininLisbon, Tired time, Tmmm, Tobby72, Tobias Conradi, Toddst1, Tombomp, Tomeasy, Tommy2010, Tomtom9041, Tone, Tony Fox, Tony1, Tonyboy27, Torch1234, Tourtux9, Toussaint, TownDown, Toyokuni3, Tpbradbury, Trachys, Trafalgar678, Travelbird, TravisTX, TreasuryTag, Treisijs, Tresiden, Tricky Victoria, Triona, Triwbe, Trusilver, Truthseeker 85.5, Tryphiodorus, Tslocum, Tto8778, Tugalese, Tugaworld, Tugazo, Turgan, Tuspm, TypoDotOrg, Tyrol5, Tyw7, Ubardak, Ufwuct, Uhhhhh777, Ukexpat, Ulmanor, Ultrauberman, Uncle Milty, Underlying lk, UniReb, Unnilquadium, Unobjectionable, Unomi, Unukorno, Urhixidur, User86654, Usgnus, Utcursch, Ute in DC, VFHenriques, VTFOHFDP, Valenciano, Van helsing, VandalCruncher, VanillaBear23, Vanished User 1004, Vanished User 8a9b4725f8376, Vardion, Vary, Vasilcho, Vasyaivanov, VaultDweller77, VbswIB2VLUBWrgbIHWFHR, VegaDark, Velella, Velho, Versus22, Victor1333, VictorO, Victorfri, Vieirasimoes, Viewfinder, Vikramkr, VincenteTorio, Violinista432, Vipinhari, Vitaleyes, Vitor cunha, Vitor107, Vitor1414, Vivio Testarossa, Vkem, Vlcmkd2, Vmts82, VolatileChemical, Volker.habl, VsevolodKrolikov, Vtrain, WTucker, WVRMad, Waltloc, Walton One, Wavelength, Wcfios, WebHamster, Welling, Welsh, Werdna, West Brom 4ever, Wetman, WhisperToMe, WikHead, Wiki alf, Wiki4share, WikiDon, WikiFew, Wikibofh, Wikitza, Wiklein, Wikopotamus, William hoste, Willrowe, Wimt, Winchelsea, Wissahickon Creek, Wj32, Wjw0111, Wknight94, Woohookitty, Wronginfosodelete, Wtmitchell, Ww2censor, Wyklety, XApple, XGenessisX, XPTO, Xandi, Xanucia, Xareu bs, Xef, Xeron220, Xgmx2, Xmendes, Xnuala, Yamaguchi, Yamamoto Ichiro, Yan Armandi, Yanksox, Ybbor, Ybnormal2day13, Ylnreddi, Ynorth, Yodaki, Yorkshirian, YourEyesOnly, Youssefsan, Yuckfoo, Yuje, Zagalejo, Zap Rowsdower, Zarhan, Zeorymer, Zeppocity, Zikander, Zoney, Zoro 25, Zotel, Zsinj, Zykon, Zzuuzz, ΚΕΚΡΩΨ, , 4760 anonymous edits

Paleontology *Source*: http://en.wikipedia.org/w/index.php?title=Paleontology *Contributors*: -jkb-, 0, 209.140.205.xxx, 28bytes, 5 albert square, A. Parrot, AJseagull1, AManWithNoPlan, AP1787, APH, AUburnTiger, Abductive, Abyssal, Acroterion, Adam78, AdjustShift, Aelffin, Ahoerstemeier, Aitias, Alansohn, Alphachimp, Altenmann, Andre Engels, Ann Stouter, Apokryltaros, Apparition11, Arbustoo, Argos'Dad, Arkuat, Arsonal, Artichoker, AshLin, Atulsnischal, Aunt Entropy, Austinbirdman, AutoGeek, Autodidactyl, AxiomShell, AzaToth, BC Museums, Ballista, Bcasterline, Bekus, Ben Hateva, Ben-Zin, Besieged, Bluezy, Bob frasier, Bobo192, Boing! said Zebedee, BookGuru, Branddobbe, Brya, Bryan Derksen, Calabe1992, Caltas, Can't sleep, clown will eat me, Catbar, Cephal-odd, Chino, Chris83, Christian75, Cmcjackson, Colonies Chris, CommonsDelinker, ConstanceScrumptious, Conversion script, Cool Blue, D.Griffon, DMacks, DabMachine, Dacaz5, Danilot, Dante Alighieri, Darkwind, Darth Panda, Daviesns, Delicious carbuncle, Dendodge, Deor, Dersonlwd, Devnullore, Dinosaur puppy, Diucón, Dlloyd, Dmerrill, Don Kenney, Doug, Dr.Bastedo, Dragon Helm, Dusik, Dysepsion, Dysmorodrepanis, E. Fokker, ESkog, Ecobio, Ed1994, EdMcMahon, El C, Enaidmawr, Enbrightenment, Enchanter, Enlil Ninlil, Epbr123, Eugenie Rowley, Eumys, Excirial, Fan-1967, Fang 23, FeralDruid, Firsfron, Flamebroil, Formeruser-81, Frankenpuppy, Freedomlinux, Fvasconcellos, G Clark, Gaelen S., Gaius Cornelius, Gary King, Gauss, Geologyguy, Giant Blue Anteater, Gill Giller Gillerger, Gogo Dodo, Graham87, Grdavies00, Greyhood, GrittyLobo441, Gsk-E, Gutsul, Hadal, Haham hanuka, Hans Dunkelberg, Haroldbethwelsh, Headbomb, Hekerui, Hmains, Hryhorash, Iridescent, J. Spencer, J.delanoy, JCPerrott, Jackol, Jagged 85, Jauhienij, Java13690, Jaysweet, JenBow, Jennavecia, JhanCRUSH, Jitterro, JohnCD, Johnny99, Jomunro, Josh3580, Journalist, Joy, Jpatrick77, Jusdafax, KFP, Kare Kare, Kevmin, Kilo-Lima, Knutux, Kokoriko, Koyaanis Qatsi, Krisfunk27, Kumioko, Les boys, Lexor, Liopleurodon93, Lklk1q2w3e4r5t, LouI, Lquilter, Luckas Blade, Luk, MER-C, MPF, Maastrictian, Mani1, Marshallsumter, Matthew Yeager, Mattjs, Maurreen, Mav, Maxamegalon2000, Mayumashu, Mchavez, Mikenorton, Minto 47, Miyagawa, Mnmngb, Mr Stephen, Mwtoews, My name, Nakon, Narayanese, NatureA16, Naudefj, NawlinWiki, Nev1, Newone, NickW, NigelR, Nihiltres, Nikai, Nishkid64, Nk, Nlyte.Software, Notorious4life, NovaDog, Nowimnthing, Numbo3, Nuno Tavares, Nwbeeson, Olcoffin, Old Father Time, Omnipedian, Oneiros, Ortolan88, Ottawa4ever, Oxymoron83, Peter.C, PeterSymonds, Philcha, Philip Trueman, PhilipO, Piano non troppo, Picus viridis, Pigsonthewing, Piloscia, Pinkadelica, Pollinator, Profberger, Quadell, RadarCzar, RandomP, Rdsmith4, Reach Out to the Truth, RedWolf, Rich Farmbrough, Richard001, Rjwilmsi, Rlinfinity, Rodasmith, RoyBoy, Rrburke, Rusty Cashman, Ruszewski, Sander123, Sankalpdravid, SchfiftyThree, Sciurinæ, Shadowlynk, Shanes, Sharonlees, Skysmith, Smallweed, Smeira, Smith609, Spiff, SpikeToronto, Spotty11222, SpuriousQ, Squidnchips, Staffwaterboy, Stefeyboy, Steinsky, Stemonitis, Stevertigo, SunCreator, Suntzu3500, Systolic, TCorp, Tagishsimon, Tassin islam, Template namespace initialisation script, The Nut, The Thing That Should Not Be, Theaucitron, Thegreyanomaly, Theirishpianist, Theman123, Thingg, Tiddly Tom, Tide rolls, Tillman, TimVickers, Tingtongbingbongboo, Tom Morris, Tom harrison, Tombomp, Tookhand, Treypat, Twinsday, Tytrain, Ukexpat, Uncle Dick, UpstateNYer, Vagary, Vardos, Veliger, Versus22, VinnytotheK, Voyagerfan5761, Vsmith, Vuong Ngan Ha, WGee, WacoJacko, Walbe470, Wavelength, Wayne Slam, Wereon, Wetman, Wiglaf, Wiki alf, WikiLaurent, Wilson44691, WolfmanSF, Wwoods, XJamRastafire, Xnuala, Yaneedalife, Yorkshirian, Zamphuor, Zandperl, Zomic13, Zundark, 562 anonymous edits

Specific name (zoology) *Source*: http://en.wikipedia.org/w/index.php?title=Specific_name_%28zoology%29 *Contributors*: Angrense, Angusmclellan, Borgx, Brya, Centrx, Circeus, Clarityfiend, D6, Dendrid, Dysmorodrepanis, Groosy, Invertzoo, Iph, Joelr31, Kaldari, Kenilworth Terrace, Kingdon, Kthejoker, Nono64, Ospalh, Peter coxhead, PigFlu Oink, Shrumster, Shyamal, Stemonitis, 9 anonymous edits

Image Sources, Licenses and Contributors

File:Red Pencil Icon.png *Source*: http://en.wikipedia.org/w/index.php?title=File:Red_Pencil_Icon.png *License*: unknown *Contributors*: User:Peter coxhead

File:Biological_classification_L_Pengo_vflip.svg *Source*: http://en.wikipedia.org/w/index.php?title=File:Biological_classification_L_Pengo_vflip.svg *License*: unknown *Contributors*: Adrignola, ArnoLagrange, Nisetpdajsankha, Pavel55, Pengo

Image:Triceratops AMNH 01.jpg *Source*: http://en.wikipedia.org/w/index.php?title=File:Triceratops_AMNH_01.jpg *License*: unknown *Contributors*: Michael Gray from Wantagh NY, USA

File:Stegosaurus at the Field Museum.jpg *Source*: http://en.wikipedia.org/w/index.php?title=File:Stegosaurus_at_the_Field_Museum.jpg *License*: unknown *Contributors*: Original uploader was Killdevil at en.wikipedia

File:Edmontonia dinosaur.png *Source*: http://en.wikipedia.org/w/index.php?title=File:Edmontonia_dinosaur.png *License*: unknown *Contributors*: user:LadyofHats

File:Sprawling and erect hip joints - horiz.png *Source*: http://en.wikipedia.org/w/index.php?title=File:Sprawling_and_erect_hip_joints_-_horiz.png *License*: unknown *Contributors*: Original uploader was Philcha at en.wikipedia (Original text : Philcha (talk))

Image:Marasuchus.JPG *Source*: http://en.wikipedia.org/w/index.php?title=File:Marasuchus.JPG *License*: unknown *Contributors*: HoopoeBaijiKite Original uploader was Mitternacht90 at en.wikipedia

Image:Herrerasaurusskeleton.jpg *Source*: http://en.wikipedia.org/w/index.php?title=File:Herrerasaurusskeleton.jpg *License*: unknown *Contributors*: Zach Tirrell from Plymouth, USA

Image:Saurischia.png *Source*: http://en.wikipedia.org/w/index.php?title=File:Saurischia.png *License*: unknown *Contributors*: FunkMonk, Koobak, Muriel Gottrop

Image:Tyrannosaurus pelvis left.jpg *Source*: http://en.wikipedia.org/w/index.php?title=File:Tyrannosaurus_pelvis_left.jpg *License*: unknown *Contributors*: User:Ballista

Image:Ornithischia.png *Source*: http://en.wikipedia.org/w/index.php?title=File:Ornithischia.png *License*: unknown *Contributors*: Koobak, Muriel Gottrop, 2 anonymous edits

Image:Edmontosaurus pelvis left.jpg *Source*: http://en.wikipedia.org/w/index.php?title=File:Edmontosaurus_pelvis_left.jpg *License*: unknown *Contributors*: Conty, Kevmin, Smeira, Timichal

Image:Macronaria scrubbed enh.jpg *Source*: http://en.wikipedia.org/w/index.php?title=File:Macronaria_scrubbed_enh.jpg *License*: unknown *Contributors*: Original uploader was Killdevil at en.wikipedia

Image:Ornithopods jconway.jpg *Source*: http://en.wikipedia.org/w/index.php?title=File:Ornithopods_jconway.jpg *License*: unknown *Contributors*: Dudo, FunkMonk, G.dallorto, John.Conway, Kevmin, Koobak

Image:Largestdinosaursbysuborder scale.png *Source*: http://en.wikipedia.org/w/index.php?title=File:Largestdinosaursbysuborder_scale.png *License*: unknown *Contributors*: User:Dinoguy2

Image:Giraffatitan scale.png *Source*: http://en.wikipedia.org/w/index.php?title=File:Giraffatitan_scale.png *License*: unknown *Contributors*: User:Dinoguy2

Image:Human-eoraptor size comparison(v2).png *Source*: http://en.wikipedia.org/w/index.php?title=File:Human-eoraptor_size_comparison(v2).png *License*: unknown *Contributors*: User:Dropzink

Image:Maiasaurusnest.jpg *Source*: http://en.wikipedia.org/w/index.php?title=File:Maiasaurusnest.jpg *License*: unknown *Contributors*: Dudo, Kevmin, Koobak

Image:Dino eggP9240092.JPG *Source*: http://en.wikipedia.org/w/index.php?title=File:Dino_eggP9240092.JPG *License*: unknown *Contributors*: user:deror_avi

Image:Centrosaurus dinosaur.png *Source*: http://en.wikipedia.org/w/index.php?title=File:Centrosaurus_dinosaur.png *License*: unknown *Contributors*: user:LadyofHats

Image:Palais de la Decouverte Tyrannosaurus rex p1050042.jpg *Source*: http://en.wikipedia.org/w/index.php?title=File:Palais_de_la_Decouverte_Tyrannosaurus_rex_p1050042.jpg *License*: unknown *Contributors*: User:David.Monniaux

Image:Eubrontes01.JPG *Source*: http://en.wikipedia.org/w/index.php?title=File:Eubrontes01.JPG *License*: unknown *Contributors*: User:Wilson44691

Image:SArchaeopteryxBerlin2.jpg *Source*: http://en.wikipedia.org/w/index.php?title=File:SArchaeopteryxBerlin2.jpg *License*: unknown *Contributors*: Owen, Richard

File:Pneumatopores on the left ilium of the theropod Aerosteon riocoloradensis.jpg *Source*: http://en.wikipedia.org/w/index.php?title=File:Pneumatopores_on_the_left_ilium_of_the_theropod_Aerosteon_riocoloradensis.jpg *License*: unknown *Contributors*: Sereno PC, Martinez RN, Wilson JA, Varricchio DJ, Alcober OA, et al.

Image:Chicxulub radar topography.jpg *Source*: http://en.wikipedia.org/w/index.php?title=File:Chicxulub_radar_topography.jpg *License*: unknown *Contributors*: NASA/JPL-Caltech

Image:Stego-marsh-1896-US geological survey.png *Source*: http://en.wikipedia.org/w/index.php?title=File:Stego-marsh-1896-US_geological_survey.png *License*: unknown *Contributors*: User:Anetode

Image:William Buckland detail.png *Source*: http://en.wikipedia.org/w/index.php?title=File:William_Buckland_detail.png *License*: unknown *Contributors*: Original uploader was Killdevil at en.wikipedia

Image:OthnielCharlesMarsh.jpg *Source*: http://en.wikipedia.org/w/index.php?title=File:OthnielCharlesMarsh.jpg *License*: unknown *Contributors*: Kilom691, Librotyrannus, Shizhao, Väsk, 1 anonymous edits

Image:edcope.jpg *Source*: http://en.wikipedia.org/w/index.php?title=File:Edcope.jpg *License*: unknown *Contributors*: F. Gutekunst UNIQ-ref-1-8950c8409189e6db-QINU

Image:LateJurassicGlobal.jpg *Source*: http://en.wikipedia.org/w/index.php?title=File:LateJurassicGlobal.jpg *License*: unknown *Contributors*: Dr. Ron Blakey - http://jan.ucc.nau.edu/~rcb7/

File:Europasaurus holgeri Scene 2.jpg *Source*: http://en.wikipedia.org/w/index.php?title=File:Europasaurus_holgeri_Scene_2.jpg *License*: unknown *Contributors*: Gerhard Boeggemann

File:MakhteshGadolCenter02.jpg *Source*: http://en.wikipedia.org/w/index.php?title=File:MakhteshGadolCenter02.jpg *License*: unknown *Contributors*: User:Wilson44691

File:MorrisonType-2.JPG *Source*: http://en.wikipedia.org/w/index.php?title=File:MorrisonType-2.JPG *License*: unknown *Contributors*: Ankyman

File:Gigandipus.JPG *Source*: http://en.wikipedia.org/w/index.php?title=File:Gigandipus.JPG *License*: unknown *Contributors*: User:Wilson44691

File:SEUtahStrat.JPG *Source*: http://en.wikipedia.org/w/index.php?title=File:SEUtahStrat.JPG *License*: unknown *Contributors*: User:Citypeek

File:Leedsi&Liopl DB.jpg *Source*: http://en.wikipedia.org/w/index.php?title=File:Leedsi&Liopl_DB.jpg *License*: unknown *Contributors*: Bogdanov dmitrchel@mail.ru

File:Fischsaurier fg01.jpg *Source*: http://en.wikipedia.org/w/index.php?title=File:Fischsaurier_fg01.jpg *License*: unknown *Contributors*: User:Dysmachus

File:Muraenosaurus l2.jpg *Source*: http://en.wikipedia.org/w/index.php?title=File:Muraenosaurus_l2.jpg *License*: unknown *Contributors*: DiBgd, Haplochromis, Putnik

File:JurassicMarineIsrael.JPG *Source*: http://en.wikipedia.org/w/index.php?title=File:JurassicMarineIsrael.JPG *License*: unknown *Contributors*: User:Wilson44691

File:SeismosaurusDB.jpg *Source*: http://en.wikipedia.org/w/index.php?title=File:SeismosaurusDB.jpg *License*: unknown *Contributors*: Dinoguy2, Haplochromis, Kevmin, 1 anonymous edits

File:Allosaurus BW.jpg *Source*: http://en.wikipedia.org/w/index.php?title=File:Allosaurus_BW.jpg *License*: unknown *Contributors*: User:ArthurWeasley

File:Stegosaurus BW.jpg *Source*: http://en.wikipedia.org/w/index.php?title=File:Stegosaurus_BW.jpg *License*: unknown *Contributors*: User:ArthurWeasley

File:Archaeopteryx lithographica (Berlin specimen).jpg *Source*: http://en.wikipedia.org/w/index.php?title=File:Archaeopteryx_lithographica_(Berlin_specimen).jpg *License*: unknown *Contributors*: User:Vesta

File:Douglas fir leaves and bud.jpg *Source*: http://en.wikipedia.org/w/index.php?title=File:Douglas_fir_leaves_and_bud.jpg *License*: unknown *Contributors*: Original uploader was UtherSRG at en.wikipedia

file:Ornithopods_jconway.jpg *Source*: http://en.wikipedia.org/w/index.php?title=File:Ornithopods_jconway.jpg *License*: unknown *Contributors*: Dudo, FunkMonk, G.dallorto, John.Conway, Kevmin, Koobak

File:Iguanodon feet.JPG *Source*: http://en.wikipedia.org/w/index.php?title=File:Iguanodon_feet.JPG *License*: unknown *Contributors*: Original uploader was Ballista at en.wikipedia

Image:Largestornithopods scale.png *Source*: http://en.wikipedia.org/w/index.php?title=File:Largestornithopods_scale.png *License*: unknown *Contributors*: User:Dinoguy2

File:Hadrosauroids.jpg *Source*: http://en.wikipedia.org/w/index.php?title=File:Hadrosauroids.jpg *License*: unknown *Contributors*: Pavel Riha = user Pavel.Riha.CB (e-mail)

File:Pt-lnh1.png *Source*: http://en.wikipedia.org/w/index.php?title=File:Pt-lnh1.png *License*: unknown *Contributors*: Brian Boru, J. Patrick Fischer, Luso-Tuga, Magul

File:LNH.png *Source*: http://en.wikipedia.org/w/index.php?title=File:LNH.png *License*: unknown *Contributors*: Brian Boru, Magul

File:LocalLourinha.svg *Source*: http://en.wikipedia.org/w/index.php?title=File:LocalLourinha.svg *License*: unknown *Contributors*: User:Rei-artur

File:Flag of Portugal.svg *Source*: http://en.wikipedia.org/w/index.php?title=File:Flag_of_Portugal.svg *License*: unknown *Contributors*: User:Nightstallion

Image:Convento de Sto. António - Lourinhã.jpg *Source*: http://en.wikipedia.org/w/index.php?title=File:Convento_de_Sto._António_-_Lourinhã.jpg *License*: unknown *Contributors*: User:Rei-artur

Image:Praia-areia-branca.jpg *Source*: http://en.wikipedia.org/w/index.php?title=File:Praia-areia-branca.jpg *License*: unknown *Contributors*: User:Love Portugal

File:coat of arms of Portugal.svg *Source*: http://en.wikipedia.org/w/index.php?title=File:Coat_of_arms_of_Portugal.svg *License*: unknown *Contributors*: User:Tonyjeff

File:EU-Portugal with islands circled.svg *Source*: http://en.wikipedia.org/w/index.php?title=File:EU-Portugal_with_islands_circled.svg *License*: unknown *Contributors*: User:NuclearVacuum, User:Pethrus

File:Increase2.svg *Source*: http://en.wikipedia.org/w/index.php?title=File:Increase2.svg *License*: unknown *Contributors*: Sarang

Image:Speakerlink.svg *Source*: http://en.wikipedia.org/w/index.php?title=File:Speakerlink.svg *License*: unknown *Contributors*: Woodstone. Original uploader was Woodstone at en.wikipedia

File:Evora-RomanTemple edit.jpg *Source*: http://en.wikipedia.org/w/index.php?title=File:Evora-RomanTemple_edit.jpg *License*: unknown *Contributors*: User:Alvesgaspar

File:Spanish reconquista.gif *Source*: http://en.wikipedia.org/w/index.php?title=File:Spanish_reconquista.gif *License*: unknown *Contributors*: Astrokey44

File:Castelo de Guimaraes.jpg *Source*: http://en.wikipedia.org/w/index.php?title=File:Castelo_de_Guimaraes.jpg *License*: unknown *Contributors*: Filipe Fortes from New York, United States

File:Vascodagama.JPG *Source*: http://en.wikipedia.org/w/index.php?title=File:Vascodagama.JPG *License*: unknown *Contributors*: AndreasPraefcke, Dantadd, Mattes, Nuno Tavares, OsamaK, Ranveig, Roland zh, Saibo, Santosga, Thib Phil, 3 anonymous edits

File:Joao IV proclaimed king.jpg *Source*: http://en.wikipedia.org/w/index.php?title=File:Joao_IV_proclaimed_king.jpg *License*: unknown *Contributors*: Abalg, Gveret Tered, Picus viridis, Santosga, Tm

File:Portugal Império total.png *Source*: http://en.wikipedia.org/w/index.php?title=File:Portugal_Império_total.png *License*: unknown *Contributors*: Original uploader was The Ogre at en.wikipedia Later versions were uploaded by The Red Hat of Pat Ferrick at en.wikipedia.

File:Louis-Michel van Loo 003.jpg *Source*: http://en.wikipedia.org/w/index.php?title=File:Louis-Michel_van_Loo_003.jpg *License*: unknown *Contributors*: User:Manuel Anastácio

File:1755 Lisbon earthquake.jpg *Source*: http://en.wikipedia.org/w/index.php?title=File:1755_Lisbon_earthquake.jpg *License*: unknown *Contributors*: Awadewit, Before My Ken, BlackIceNRW, Chris 73, Dereckson, EuTuga, Interpretix, Jaranda, Kingruedi, Lusitana, 1 anonymous edits

File:Príncipe Regente de Portugal e toda a Família Real embarcando para Brasil no cais de Belém.jpg *Source*: http://en.wikipedia.org/w/index.php?title=File:Príncipe_Regente_de_Portugal_e_toda_a_Família_Real_embarcando_para_Brasil_no_cais_de_Belém.jpg *License*: unknown *Contributors*: Henry L'Évêque (1768-1845)

File:Brazil states1823.png *Source*: http://en.wikipedia.org/w/index.php?title=File:Brazil_states1823.png *License*: unknown *Contributors*: Electionworld, Pedro Aguiar, Quissamã, Raphael.lorenzeto, 3 anonymous edits

File:Mapa Cor-de-Rosa.svg *Source*: http://en.wikipedia.org/w/index.php?title=File:Mapa_Cor-de-Rosa.svg *License*: unknown *Contributors*: User:Lokal_Profil

File:35- Rei D. Manuel II - O Patriota.jpg *Source*: http://en.wikipedia.org/w/index.php?title=File:35-_Rei_D._Manuel_II_-_O_Patriota.jpg *License*: unknown *Contributors*: User L Gonçalves on pt.wikipedia

File:Portuguese colonial war map1.PNG *Source*: http://en.wikipedia.org/w/index.php?title=File:Portuguese_colonial_war_map1.PNG *License*: unknown *Contributors*: Bourrichon, Hoshie, Ms2ger, Nuno Tavares, The Ogre

File:Tratado de Lisboa 13 12 2007 (081).jpg *Source*: http://en.wikipedia.org/w/index.php?title=File:Tratado_de_Lisboa_13_12_2007_(081).jpg *License*: unknown *Contributors*: Archiwum Kancelarii Prezydenta RP

File:Pico-HortaFaial.jpg *Source*: http://en.wikipedia.org/w/index.php?title=File:Pico-HortaFaial.jpg *License*: unknown *Contributors*: antlewis

File:Lagoa das Sete Cidades, Miradouro da vista do Rei, Sete Cidades, Ponta Delgada, ilha de São Miguel, Açores.JPG *Source*: http://en.wikipedia.org/w/index.php?title=File:Lagoa_das_Sete_Cidades,_Miradouro_da_vista_do_Rei,_Sete_Cidades,_Ponta_Delgada,_ilha_de_São_Miguel,_Açores.JPG *License*: unknown *Contributors*: User:Luissilveira

File:Alentejo oak on wheat field.jpg *Source*: http://en.wikipedia.org/w/index.php?title=File:Alentejo_oak_on_wheat_field.jpg *License*: unknown *Contributors*: AnRo0002, C90, FlickrLickr, FlickreviewR, Ilmari Karonen, Lusitana, MPF, 1 anonymous edits

File:Dona Ana beach, Lagos.jpg *Source*: http://en.wikipedia.org/w/index.php?title=File:Dona_Ana_beach,_Lagos.jpg *License*: unknown *Contributors*: User:Lacobrigo

File:Estrela Março 2010-16b.jpg *Source*: http://en.wikipedia.org/w/index.php?title=File:Estrela_Março_2010-16b.jpg *License*: unknown *Contributors*: User:Alvesgaspar

File:Régua Douro 4.JPG *Source*: http://en.wikipedia.org/w/index.php?title=File:Régua_Douro_4.JPG *License*: unknown *Contributors*: User:Husond

File:Cavaco Silva 2007.jpg *Source*: http://en.wikipedia.org/w/index.php?title=File:Cavaco_Silva_2007.jpg *License*: unknown *Contributors*: Ricardo Stuckert/PR

File:Flickr - europeanpeoplesparty - EPP Summit June 2010 (82).jpg *Source*: http://en.wikipedia.org/w/index.php?title=File:Flickr_-_europeanpeoplesparty_-_EPP_Summit_June_2010_(82).jpg *License*: unknown *Contributors*: European People's Party

File:PortugalNumbered.png *Source*: http://en.wikipedia.org/w/index.php?title=File:PortugalNumbered.png *License*: unknown *Contributors*: Mortadelo2005

File:Olivença location.PNG *Source*: http://en.wikipedia.org/w/index.php?title=File:Olivença_location.PNG *License*: unknown *Contributors*: User:Jsobral

File:F16A FAP refuel KC-10.jpg *Source*: http://en.wikipedia.org/w/index.php?title=File:F16A_FAP_refuel_KC-10.jpg *License*: unknown *Contributors*: Bouwe Brouwer, Cobatfor, Get It, Morio, Tm

File:POS Corte Real (F 332).jpg *Source*: http://en.wikipedia.org/w/index.php?title=File:POS_Corte_Real_(F_332).jpg *License*: unknown *Contributors*: MMass Communication Specialist Seaman Apprentice Shonna Cunningham

File:Maia Camara Municipal.jpg *Source*: http://en.wikipedia.org/w/index.php?title=File:Maia_Camara_Municipal.jpg *License*: unknown *Contributors*: User:Pedrovsk

Image:Obidos 10.JPG *Source*: http://en.wikipedia.org/w/index.php?title=File:Obidos_10.JPG *License*: unknown *Contributors*: Lusitana, OsvaldoGago, Tm

Image:Vilamoura.jpg *Source*: http://en.wikipedia.org/w/index.php?title=File:Vilamoura.jpg *License*: unknown *Contributors*: user:Valter Jacinto

File:LGPK.jpg *Source*: http://en.wikipedia.org/w/index.php?title=File:LGPK.jpg *License*: unknown *Contributors*: Fred mendonca at en.wikipedia

File:Pena National Palace.JPG *Source*: http://en.wikipedia.org/w/index.php?title=File:Pena_National_Palace.JPG *License*: unknown *Contributors*: User:Guillaume70

File:Almourol 034.jpg *Source*: http://en.wikipedia.org/w/index.php?title=File:Almourol_034.jpg *License*: unknown *Contributors*: Alvaro, Daniel Feliciano, Lusitana

File:Metro do Porto Flexity Outlook Eurotram Trindade.jpg *Source*: http://en.wikipedia.org/w/index.php?title=File:Metro_do_Porto_Flexity_Outlook_Eurotram_Trindade.jpg *License*: unknown *Contributors*: User:Jcornelius

File:IMG 0224.JPG *Source*: http://en.wikipedia.org/w/index.php?title=File:IMG_0224.JPG *License*: unknown *Contributors*: -

File:Vasco de Gama bridge side.jpg *Source*: http://en.wikipedia.org/w/index.php?title=File:Vasco_de_Gama_bridge_side.jpg *License*: unknown *Contributors*: FlickrLickr, FlickreviewR, Herbythyme, Para, Ronaldino

File:Ponte25Abril1.jpg *Source*: http://en.wikipedia.org/w/index.php?title=File:Ponte25Abril1.jpg *License*: unknown *Contributors*: Vitor Oliveira from Torres Vedras, PORTUGAL

File:up reitoria.jpg *Source*: http://en.wikipedia.org/w/index.php?title=File:Up_reitoria.jpg *License*: unknown *Contributors*: User:UP

File:Moniz.jpg *Source*: http://en.wikipedia.org/w/index.php?title=File:Moniz.jpg *License*: unknown *Contributors*: Ephraim33, JotaCartas, LX, Synaptics

File:Lisboa-Oceanario.jpg *Source*: http://en.wikipedia.org/w/index.php?title=File:Lisboa-Oceanario.jpg *License*: unknown *Contributors*: Juntas, Nol Aders, Para, Pristigaster

File:Alqueva dam-Rio Guadiana.JPG *Source*: http://en.wikipedia.org/w/index.php?title=File:Alqueva_dam-Rio_Guadiana.JPG *License*: unknown *Contributors*: Hugo.arg, Tm

File:SolarPowerPlantSerpa.jpg *Source*: http://en.wikipedia.org/w/index.php?title=File:SolarPowerPlantSerpa.jpg *License*: unknown *Contributors*: Original uploader was user:Ceinturion at en.wikipedia

File:Population density of Portugal.PNG *Source*: http://en.wikipedia.org/w/index.php?title=File:Population_density_of_Portugal.PNG *License*: unknown *Contributors*: User:Kjetil r, User:Susana Freixeiro

File:Santuario di Fatima 2.JPG *Source*: http://en.wikipedia.org/w/index.php?title=File:Santuario_di_Fatima_2.JPG *License*: unknown *Contributors*: User:Aracuano

File:Lusophone World.svg *Source*: http://en.wikipedia.org/w/index.php?title=File:Lusophone_World.svg *License*: unknown *Contributors*: User:Saftorangen

File:UNL presidency building.jpg *Source*: http://en.wikipedia.org/w/index.php?title=File:UNL_presidency_building.jpg *License*: unknown *Contributors*: Joaonsduarte

File:Coimbra University Tower 2.jpg *Source*: http://en.wikipedia.org/w/index.php?title=File:Coimbra_University_Tower_2.jpg *License*: unknown *Contributors*: Original uploader was Elapsed at en.wikipedia Later version(s) were uploaded by Arpingstone at en.wikipedia.

File:Hospital de Santa Maria.jpg *Source*: http://en.wikipedia.org/w/index.php?title=File:Hospital_de_Santa_Maria.jpg *License*: unknown *Contributors*: User:Ivendrell

File:Torre Belém April 2009-4a.jpg *Source*: http://en.wikipedia.org/w/index.php?title=File:Torre_Belém_April_2009-4a.jpg *License*: unknown *Contributors*: User:Alvesgaspar

File:Luís de Camões por François Gérard.jpg *Source*: http://en.wikipedia.org/w/index.php?title=File:Luís_de_Camões_por_François_Gérard.jpg *License*: unknown *Contributors*: CommonsDelinker, Dantadd, LX

File:Bacalhau a Bras.jpg *Source*: http://en.wikipedia.org/w/index.php?title=File:Bacalhau_a_Bras.jpg *License*: unknown *Contributors*: User:Fpenteado

File:Jose malhoa fado.jpg *Source*: http://en.wikipedia.org/w/index.php?title=File:Jose_malhoa_fado.jpg *License*: unknown *Contributors*: Dcoetzee, Lalupa, Rei-artur

File:Casamusicaexterior.jpg *Source*: http://en.wikipedia.org/w/index.php?title=File:Casamusicaexterior.jpg *License*: unknown *Contributors*: User:Marinhopaiva

File:Euro2004OpeningCeremony.jpg *Source*: http://en.wikipedia.org/w/index.php?title=File:Euro2004OpeningCeremony.jpg *License*: unknown *Contributors*: Joaocastro (Joao Castro) at en.wikipedia

File:Parque das Na aes Pavilh o Atl ntico Foto Lu s Ponte 1-1-.jpg *Source*: http://en.wikipedia.org/w/index.php?title=File:Parque_das_Na_aes_Pavilh_o_Atl_ntico_Foto_Lu_s_Ponte_1-1-.jpg *License*: unknown *Contributors*: cm lisboa

File:Bodyboardingpauldomar2.JPG *Source*: http://en.wikipedia.org/w/index.php?title=File:Bodyboardingpauldomar2.JPG *License*: unknown *Contributors*: User:Abbiati2

File:Geological time spiral - sharper.png *Source*: http://en.wikipedia.org/w/index.php?title=File:Geological_time_spiral_-_sharper.png *License*: unknown *Contributors*: United States Geological Survey. Original uploader was Philcha at en.wikipedia

File:Joda paleontologist.jpg *Source*: http://en.wikipedia.org/w/index.php?title=File:Joda_paleontologist.jpg *License*: unknown *Contributors*: NPS

File:Europasaurus Praeparation.JPG *Source*: http://en.wikipedia.org/w/index.php?title=File:Europasaurus_Praeparation.JPG *License*: unknown *Contributors*: Nils Knötschke

File:Paleontologist chipping.jpg *Source*: http://en.wikipedia.org/w/index.php?title=File:Paleontologist_chipping.jpg *License*: unknown *Contributors*: FunkMonk, Pieter Kuiper, Quadell

File:Palais de la Decouverte Tyrannosaurus rex p1050042.jpg *Source*: http://en.wikipedia.org/w/index.php?title=File:Palais_de_la_Decouverte_Tyrannosaurus_rex_p1050042.jpg *License*: unknown *Contributors*: User:David.Monniaux

File:LateCarboniferousGlobal.jpg *Source*: http://en.wikipedia.org/w/index.php?title=File:LateCarboniferousGlobal.jpg *License*: unknown *Contributors*: Dr. Ron Blakey - http://jan.ucc.nau.edu/~rcb7/

File:Marrella_(fossil).png *Source*: http://en.wikipedia.org/w/index.php?title=File:Marrella_(fossil).png *License*: unknown *Contributors*: Original uploader was Verisimilus at en.wikipedia

File:CambrianRusophycus.jpg *Source*: http://en.wikipedia.org/w/index.php?title=File:CambrianRusophycus.jpg *License*: unknown *Contributors*: User:Wilson44691

File:Biological classification L Pengo.svg *Source*: http://en.wikipedia.org/w/index.php?title=File:Biological_classification_L_Pengo.svg *License*: unknown *Contributors*: Peter Halasz. (w:User:PengoUser:Pengo)

Image:Index fossils blank 01.png *Source*: http://en.wikipedia.org/w/index.php?title=File:Index_fossils_blank_01.png *License*: unknown *Contributors*: United States Geological Survey. Original uploader was Philcha at en.wikipedia

File:Runzelmarken.jpg *Source*: http://en.wikipedia.org/w/index.php?title=File:Runzelmarken.jpg *License*: unknown *Contributors*: User:Smith609

File:Opabinia BW2.jpg *Source*: http://en.wikipedia.org/w/index.php?title=File:Opabinia_BW2.jpg *License*: unknown *Contributors*: User:ArthurWeasley

File:Yanoconodon BW.jpg *Source*: http://en.wikipedia.org/w/index.php?title=File:Yanoconodon_BW.jpg *License*: unknown *Contributors*: User:ArthurWeasley

File:House sparrow04.jpg *Source*: http://en.wikipedia.org/w/index.php?title=File:House_sparrow04.jpg *License*: unknown *Contributors*: User:Fir0002

File:bee pollenating a rose.jpg *Source*: http://en.wikipedia.org/w/index.php?title=File:Bee_pollenating_a_rose.jpg *License*: unknown *Contributors*: Kaldari, Lalupa, Satyrium, UpstateNYer, Wofl

Image:Extinction intensity.svg *Source*: http://en.wikipedia.org/w/index.php?title=File:Extinction_intensity.svg *License*: unknown *Contributors*: Beland, Dragons flight, Geoffrey.landis, Patrick, Smith609, TomCatX, Wst, 1 anonymous edits

Image:Phanerozoic biodiversity blank 01.png *Source*: http://en.wikipedia.org/w/index.php?title=File:Phanerozoic_biodiversity_blank_01.png *License*: unknown *Contributors*: User:Philcha

File:Cuvier elephant jaw.jpg *Source*: http://en.wikipedia.org/w/index.php?title=File:Cuvier_elephant_jaw.jpg *License*: unknown *Contributors*: w:Georges CuvierGeorges Cuvier (1769–1832)

File:Haikouichthys4.png *Source*: http://en.wikipedia.org/w/index.php?title=File:Haikouichthys4.png *License*: unknown *Contributors*: User:Giant Blue Anteater

GNU Free Documentation License Version 1.2, November 2002 Copyright (C) 2000,2001,2002 Free Software Foundation, Inc. 59 Temple Place, Suite 330, Boston, MA 02111-1307 USA Everyone is permitted to copy and distribute verbatim copies of this license document, but changing it is not allowed.

0. PREAMBLE

The purpose of this License is to make a manual, textbook, or other functional and useful document "free" in the sense of freedom: to assure everyone the effective freedom to copy and redistribute it, with or without modifying it, either commercially or noncommercially. Secondarily, this License preserves for the author and publisher a way to get credit for their work, while not being considered responsible for modifications made by others. This License is a kind of "copyleft", which means that derivative works of the document must themselves be free in the same sense. It complements the GNU General Public License, which is a copyleft license designed for free software. We have designed this License in order to use it for manuals for free software, because free software needs free documentation: a free program should come with manuals providing the same freedoms that the software does. But this License is not limited to software manuals; it can be used for any textual work, regardless of subject matter or whether it is published as a printed book. We recommend this License principally for works whose purpose is instruction or reference.

1. APPLICABILITY AND DEFINITIONS

This License applies to any manual or other work, in any medium, that contains a notice placed by the copyright holder saying it can be distributed under the terms of this License. Such a notice grants a world-wide, royalty-free license, unlimited in duration, to use that work under the conditions stated herein. The "Document", below, refers to any such manual or work. Any member of the public is a licensee, and is addressed as "you". You accept the license if you copy, modify or distribute the work in a way requiring permission under copyright law. A "Modified Version" of the Document means any work containing the Document or a portion of it, either copied verbatim, or with modifications and/or translated into another language. A "Secondary Section" is a named appendix or a front-matter section of the Document that deals exclusively with the relationship of the publishers or authors of the Document to the Document's overall subject (or to related matters) and contains nothing that could fall directly within that overall subject. (Thus, if the Document is in part a textbook of mathematics, a Secondary Section may not explain any mathematics.) The relationship could be a matter of historical connection with the subject or with related matters, or of legal, commercial, philosophical, ethical or political position regarding them. The "Invariant Sections" are certain Secondary Sections whose titles are designated, as being those of Invariant Sections, in the notice that says that the Document is released under this License. If a section does not fit the above definition of Secondary then it is not allowed to be designated as Invariant. The Document may contain zero Invariant Sections. If the Document does not identify any Invariant Sections then there are none. The "Cover Texts" are certain short passages of text that are listed, as Front-Cover Texts or Back-Cover Texts, in the notice that says that the Document is released under this License. A Front-Cover Text may be at most 5 words, and a Back-Cover Text may be at most 25 words. A "Transparent" copy of the Document means a machine-readable copy, represented in a format whose specification is available to the general public, that is suitable for revising the document straightforwardly with generic text editors or (for images composed of pixels) generic paint programs or (for drawings) some widely available drawing editor, and that is suitable for input to text formatters or for automatic translation to a variety of formats suitable for input to text formatters. A copy made in an otherwise Transparent file format whose markup, or absence of markup, has been arranged to thwart or discourage subsequent modification by readers is not Transparent. An image format is not Transparent if used for any substantial amount of text. A copy that is not "Transparent" is called "Opaque". Examples of suitable formats for Transparent copies include plain ASCII without markup, Texinfo input format, LaTeX input format, SGML or XML using a publicly available DTD, and standard-conforming simple HTML, PostScript or PDF designed for human modification. Examples of transparent image formats include PNG, XCF and JPG. Opaque formats include proprietary formats that can be read and edited only by proprietary word processors, SGML or XML for which the DTD and/or processing tools are not generally available, and the machine-generated HTML, PostScript or PDF produced by some word processors for output purposes only. The "Title Page" means, for a printed book, the title page itself, plus such following pages as are needed to hold, legibly, the material this License requires to appear in the title page. For works in formats which do not have any title page as such, "Title Page" means the text near the most prominent appearance of the work's title, preceding the beginning of the body of the text. A section "Entitled XYZ" means a named subunit of the Document whose title either is precisely XYZ or contains XYZ in parentheses following text that translates XYZ in another language. (Here XYZ stands for a specific section name mentioned below, such as "Acknowledgements", "Dedications", "Endorsements", or "History".) To "Preserve the Title" of such a section when you modify the Document means that it remains a section "Entitled XYZ" according to this definition. The Document may include Warranty Disclaimers next to the notice which states that this License applies to the Document. These Warranty Disclaimers are considered to be included by reference in this License, but only as regards disclaiming warranties: any other implication that these Warranty Disclaimers may have is void and has no effect on the meaning of this License.

2. VERBATIM COPYING

You may copy and distribute the Document in any medium, either commercially or noncommercially, provided that this License, the copyright notices, and the license notice saying this License applies to the Document are reproduced in all copies, and that you add no other conditions whatsoever to those of this License. You may not use technical measures to obstruct or control the reading or further copying of the copies you make or distribute. However, you may accept compensation in exchange for copies. If you distribute a large enough number of copies you must also follow the conditions in section 3. You may also lend copies, under the same conditions stated above, and you may publicly display copies.

3. COPYING IN QUANTITY

If you publish printed copies (or copies in media that commonly have printed covers) of the Document, numbering more than 100, and the Document's license notice requires Cover Texts, you must enclose the copies in covers that carry, clearly and legibly, all these Cover Texts: Front-Cover Texts on the front cover, and Back-Cover Texts on the back cover. Both covers must also clearly and legibly identify you as the publisher of these copies. The front cover must present the full title with all words of the title equally prominent and visible. You may add other material on the covers in addition. Copying with changes limited to the covers, as long as they preserve the title of the Document and satisfy these conditions, can be treated as verbatim copying in other respects. If the required texts for either cover are too voluminous to fit legibly, you should put the first ones listed (as many as fit reasonably) on the actual cover, and continue the rest onto adjacent pages. If you publish or distribute Opaque copies of the Document numbering more than 100, you must either include a machine-readable Transparent copy along with each Opaque copy, or state in or with each Opaque copy a computer-network location from which the general network-using public has access to download using public-standard network protocols a complete Transparent copy of the Document, free of added material. If you use the latter option, you must take reasonably prudent steps, when you begin distribution of Opaque copies in quantity, to ensure that this Transparent copy will remain thus accessible at the stated location until at least one year after the last time you distribute an Opaque copy (directly or through your agents or retailers) of that edition to the public. It is requested, but not required, that you contact the authors of the Document well before redistributing any large number of copies, to give them a chance to provide you with an updated version of the Document.

4. MODIFICATIONS

You may copy and distribute a Modified Version of the Document under the conditions of sections 2 and 3 above, provided that you release the Modified Version under precisely this License, with the Modified Version filling the role of the Document, thus licensing distribution and modification of the Modified Version to whoever possesses a copy of it. In addition, you must do these things in the Modified Version: A. Use in the Title Page (and on the covers, if any) a title distinct from that of the Document, and from those of previous versions (which should, if there were any, be listed in the History section of the Document). You may use the same title as a previous version if the original publisher of that version gives permission. B. List on the Title Page, as authors, one or more persons or entities responsible for authorship of the modifications in the Modified Version, together with at least five of the principal authors of the Document (all of its principal authors, if it has fewer than five), unless they release you from this requirement. C. State on the Title page the name of the publisher of the Modified Version, as the publisher. D. Preserve all the copyright notices of the Document. E. Add an appropriate copyright notice for your modifications adjacent to the other copyright notices. F. Include, immediately after the copyright notices, a license notice giving the public permission to use the Modified Version under the terms of this License, in the form shown in the Addendum below. G. Preserve in that license notice the full lists of Invariant Sections and required Cover Texts given in the Document's license notice. H. Include an unaltered copy of this License. I. Preserve the section Entitled "History", Preserve its Title, and add to it an item stating at least the title, year, new authors, and publisher of the Modified Version as given on the Title Page. If there is no section Entitled "History" in the Document, create one stating the title, year, authors, and publisher of the Document as given on its Title Page, then add an item describing the Modified Version as stated in the previous sentence. J. Preserve the network location, if any, given in the Document for public access to a Transparent copy of the Document, and likewise the network locations given in the Document for previous versions it was based on. These may be placed in the "History" section. You may omit a network location for a work that was published at least four years before the Document itself, or if the original publisher of the version it refers to gives permission. K. For any section Entitled "Acknowledgements" or "Dedications", Preserve the Title of the section, and preserve in the section all the substance and tone of each of the contributor acknowledgements and/or dedications given therein. L. Preserve all the Invariant Sections of the Document, unaltered in their text and in their titles. Section numbers or the equivalent are not considered part of the section titles. M. Delete any section Entitled "Endorsements". Such a section may not be included in the Modified Version. N. Do not retitle any existing section to be Entitled "Endorsements" or to conflict in title with any Invariant Section. O. Preserve any Warranty Disclaimers. If the Modified Version includes new front-matter sections or appendices that qualify as Secondary Sections and contain no material copied from the Document, you may at your option designate some or all of these sections as invariant. To do this, add their titles to the list of Invariant Sections in the Modified Version's license notice. These titles must be distinct from any other section titles. You may add a section Entitled "Endorsements", provided it contains nothing but endorsements of your Modified Version by various parties--for example, statements of peer review or that the text has been approved by an organization as the authoritative definition of a standard. You may add a passage of up to five words as a Front-Cover Text, and a passage of up to 25 words as a Back-Cover Text, to the end of the list of Cover Texts in the Modified Version. Only one passage of Front-Cover Text and one of Back-Cover Text may be added by (or through arrangements made by) any one entity. If the Document already includes a cover text for the same cover, previously added by you or by arrangement made by the same entity you are acting on behalf of, you may not add another; but you may replace the old one, on explicit permission from the previous publisher that added the old one. The author(s) and publisher(s) of the Document do not by this License give permission to use their names for publicity for or to assert or imply endorsement of any Modified Version.

5. COMBINING DOCUMENTS

You may combine the Document with other documents released under this License, under the terms defined in section 4 above for modified versions, provided that you include in the combination all of the Invariant Sections of all of the original documents, unmodified, and list them all as Invariant Sections of your combined work in its license notice, and that you preserve all their Warranty Disclaimers. The combined work need only contain one copy of this License, and multiple identical Invariant Sections may be replaced with a single copy. If there are multiple Invariant Sections with the same name but different contents, make the title of each such section unique by adding at the end of it, in parentheses, the name of the original author or publisher of that section if known, or else a unique number. Make the same adjustment to the section titles in the list of Invariant Sections in the license notice of the combined work. In the combination, you must combine any sections Entitled "History" in the various original documents, forming one section Entitled "History"; likewise combine any sections Entitled "Acknowledgements", and any sections Entitled "Dedications". You must delete all sections Entitled "Endorsements".

6. COLLECTIONS OF DOCUMENTS

You may make a collection consisting of the Document and other documents released under this License, and replace the individual copies of this License in the various documents with a single copy that is included in the collection, provided that you follow the rules of this License for verbatim copying of each of the documents in all other respects. You may extract a single document from such a collection, and distribute it individually under this License, provided you insert a copy of this License into the extracted document, and follow this License in all other respects regarding verbatim copying of that document.

7. AGGREGATION WITH INDEPENDENT WORKS

A compilation of the Document or its derivatives with other separate and independent documents or works, in or on a volume of a storage or distribution medium, is called an "aggregate" if the copyright resulting from the compilation is not used to limit the legal rights of the compilation's users beyond what the individual works permit. When the Document is included in an aggregate, this License does not apply to the other works in the aggregate which are not themselves derivative works of the Document. If the Cover Text requirement of section 3 is applicable to these copies of the Document, then if the Document is less than one half of the entire aggregate, the Document's Cover Texts may be placed on covers that bracket the Document within the aggregate, or the electronic equivalent of covers if the Document is in electronic form. Otherwise they must appear on printed covers that bracket the whole aggregate.

8. TRANSLATION

Translation is considered a kind of modification, so you may distribute translations of the Document under the terms of section 4. Replacing Invariant Sections with translations requires special permission from their copyright holders, but you may include translations of some or all Invariant Sections in addition to the original versions of these Invariant Sections. You may include a translation of this License, and all the license notices in the Document, and any Warranty Disclaimers, provided that you also include the original English version of this License and the original versions of those notices and disclaimers. In case of a disagreement between the translation and the original version of this License or a notice or disclaimer, the original version will prevail. If a section in the Document is Entitled "Acknowledgements", "Dedications", or "History", the requirement (section 4) to Preserve its Title (section 1) will typically require changing the actual title.

9. TERMINATION

You may not copy, modify, sublicense, or distribute the Document except as expressly provided for under this License. Any other attempt to copy, modify, sublicense or distribute the Document is void, and will automatically terminate your rights under this License. However, parties who have received copies, or rights, from you under this License will not have their licenses terminated so long as such parties remain in full compliance.

10. FUTURE REVISIONS OF THIS LICENSE

The Free Software Foundation may publish new, revised versions of the GNU Free Documentation License from time to time. Such new versions will be similar in spirit to the present version, but may differ in detail to address new problems or concerns. See http://www.gnu.org/copyleft/. Each version of the License is given a distinguishing version number. If the Document specifies that a particular numbered version of this License "or any later version" applies to it, you have the option of following the terms and conditions either of that specified version or of any later version that has been published (not as a draft) by the Free Software Foundation. If the Document does not specify a version number of this License, you may choose any version ever published (not as a draft) by the Free Software Foundation. ADDENDUM: How to use this License for your documents To use this License in a document you have written, include a copy of the License in the document and put the following copyright and license notices just after the title page: Copyright (c) YEAR YOUR NAME. Permission is granted to copy, distribute and/or modify this document under the terms of the GNU Free Documentation License, Version 1.2 or any later version published by the Free Software Foundation; with no Invariant Sections, no Front-Cover Texts, and no Back-Cover Texts. A copy of the license is included in the section entitled "GNU Free Documentation License". If you have Invariant Sections, Front-Cover Texts and Back-Cover Texts, replace the "with...Texts." line with this: with the Invariant Sections being LIST THEIR TITLES, with the Front-Cover Texts being LIST, and with the Back-Cover Texts being LIST. If you have Invariant Sections without Cover Texts, or some other combination of the three, merge those two alternatives to suit the situation. If your document contains nontrivial examples of program code, we recommend releasing these examples in parallel under your choice of free software license, such as the GNU General Public License, to permit their use in free software.

MIX
Papier aus verantwortungsvollen Quellen
Paper from responsible sources
FSC® C105338

Printed by Books on Demand GmbH, Norderstedt / Germany